Resilience and Sustainability in Architecture and Urban Planning: Policies, Practices, Strategies and Visions

Resilience and Sustainability in Architecture and Urban Planning: Policies, Practices, Strategies and Visions

Editors

Iftekhar Ahmed
Masa Noguchi
David O'Brien
Chris Tucker
Mittul Vahanvati

MDPI • Basel • Beijing • Wuhan • Barcelona • Belgrade • Manchester • Tokyo • Cluj • Tianjin

Editors

Iftekhar Ahmed
School of Architecture and
Built Environment
University of Newcastle
Callaghan
Australia

Masa Noguchi
Faculty of Architecture,
Building and Planning
University of Melbourne
Parkville
Australia

David O'Brien
Faculty of Architecture,
Building and Planning
University of Melbourne
Parkville
Australia

Chris Tucker
School of Architecture and
Built Environment
University of Newcastle
Callaghan
Australia

Mittul Vahanvati
School of Global, Urban and
Social Studies
RMIT University
Melbourne
Australia

Editorial Office
MDPI
St. Alban-Anlage 66
4052 Basel, Switzerland

This is a reprint of articles from the Special Issue published online in the open access journal *Architecture* (ISSN 2673-8945) (available at: www.mdpi.com/journal/architecture/special_issues/ resilience_sustainability_architecture).

For citation purposes, cite each article independently as indicated on the article page online and as indicated below:

LastName, A.A.; LastName, B.B.; LastName, C.C. Article Title. *Journal Name* **Year**, *Volume Number*, Page Range.

ISBN 978-3-0365-6419-7 (Hbk)
ISBN 978-3-0365-6418-0 (PDF)

Cover image courtesy of Iftekhar Ahmed

Contents

About the Editors

Iftekhar Ahmed

Dr Iftekhar Ahmed is an Associate Professor in Construction Management and Disaster Resilience, School of Architecture and Built Environment within the College of Engineering, Science and Environment. He was the former Program Convenor of the Master of Disaster Resilience and Sustainable Development, University of Newcastle, Australia. His teaching focuses on policy and social dimensions of disaster risk reduction, disaster resilience and management in the built environment and the UN's sustainable development goals. He conducts extensive research on disaster resilience, climate change adaptation, participatory development and urbanisation in the Asia-Pacific region. Dr Ahmed has written several books, professional reports and many peer-reviewed publications. Recently, he was the lead author of the book 'Disaster Resilience in South Asia'(Routledge UK, 2020). Dr Ahmed completed his Ph.D. at Oxford Brookes University, UK, Master of Science from the Massachusetts Institute of Technology (MIT), USA, and Bachelor of Architecture (Honours) from the Indian Institute of Technology (IIT), India. In the past, he worked as a project manager at the Asian Disaster Preparedness Center (ADPC), Thailand, and as a shelter specialist at the United Nations Development Programme (UNDP), Bangladesh, and taught at the Bangladesh University of Engineering and Technology (BUET). Dr Ahmed serves widely as a consultant and technical advisor for international development and humanitarian agencies and an Editorial Board Member of a number of international journals including *Architecture*.

Masa Noguchi

Dr Masa Noguchi is an Associate Professor in Environmental Design at the Faculty of Architecture, Building and Planning, University of Melbourne, Australia. He is a Chartered Engineer, Environmentalist, and Technological Product Designer registered, respectively, with the Engineering Council, Society for the Environment, and the Institution of Engineering Designers in the UK. In 2002, he also became a member of the Royal Architectural Institute of Canada, and today he serves as a Certified Passive House Designer registered with the Passive House Institute in Germany. Dr Noguchi is the founding global coordinator of ZEMCH (Zero Energy Mass Custom Home) Network, which consists of over 800 partners from nearly 40 countries and initiated several industry-academia knowledge transfer events. In ABP, he leads a ZEMCH Sustainable Design subject within the postgraduate program. Before coming to Melbourne, Dr Noguchi was a Reader at the Mackintosh School of Architecture, The Glasgow School of Art affiliated with the University of Glasgow, where he established a ZEMCH pathway within the Master of Architectural Studies program. His research is based on ZEMCH engineering design towards socially, economically, environmentally, and humanly sustainable housing unit and community developments in developed and developing countries. Serving as the Editor-in-Chief or Editorial Board Member of nearly 20 academic journals, Dr Noguchi is frequently invited to deliver keynote lectures on ZEMCH R&D global actions at national and international conferences.

David O'Brien

Dr David O'Brien is an Associate Professor at the Melbourne School of Design under the Faculty of Architecture, Building and Planning, University of Melbourne, Australia. He is the director of the award-winning Bower Studio program—internationally recognised for its innovative work co-designing and co-building community infrastructure with remote, marginalised indigenous communities in the Northern Territory, New South Wales, Western Australia, and internationally in Papua New Guinea and Thailand. This program relies on innovative pedagogical and engagement processes to form positive relationships to develop tangible benefits to communities. Outcomes have included a range of culture, health, education, and arts facilities that have been developed alongside community groups, government agencies, aid workers, industry partners, engineers, and sociologists. Dr O'Brien's academic publications are focused on links between housing cultures and technology with a specific focus on mass-produced housing developed in the aftermath of disasters. He has a particular interest is the way residents self-modify this housing to suit their own aspirational and functional needs.

Chris Tucker

Dr Chris Tucker was the Head of Discipline (Architecture) within the School of Architecture and Built Environment, now under the College of Engineering, Science and Environment, University of Newcastle, Australia. He is also the Principal of the architectural practice, HERD architects. Dr Tucker has been awarded regional, state and international prizes for architecture, and his buildings and designs have been exhibited and published locally and internationally. Working between architectural practice and research, academic projects are broadly interested in social justice, our relationship to Country, minimalism, the public capacity of the street, and the needling of the privileged but vague and generic qualities of contemporary public space. Dr Tucker also runs architectural studios in Tokyo (from 2019) and the Town Camps of Alice Springs (from 2017), and collaborates with local government and First Nation communities.

Mittul Vahanvati

Dr Mittul Vahanvati is a Senior Lecturer in sustainability and urban planning at the School of Global, Urban and Social Studies under the College of Design and Social Context, RMIT University, Australia. Dr Vahanvati leads the design and development of a suite of Urban Design courses, core to the Planning Institute of Australia (PIA) accreditation. She co-leads two research groups—Climate Change Transformations Research Group and Future Urban Researchers Network. Dr Vahanvati is also a co-founder of a Melbourne based design studio and social enterprise—Giant Grass. Her research focuses on climate change and disaster resilience of housing and communities in the Asia-Pacific region. Dr Vahanvati's's research adopts a co-design and co-production approach, whereby she works with people living in informal settlements and vulnerability. She engages with UN bodies and national and local governments in Australia and the Pacific, as well as NGOs. Dr Vahanvati completed her Ph.D. at RMIT University, Australia, Master of Sustainable Development in the Built environment from University of New South Wales, Sydney, Australian and Bachelor of Architecture (Honours) from Centre for Environmental Planning and Technology, Ahmedabad, India. Before moving into academia, Dr Vahanvati practiced in architecture and sustainability industry for eight years, across India, Switzerland, and Australia.

Preface to "Resilience and Sustainability in Architecture and Urban Planning: Policies, Practices, Strategies and Visions"

The production and development of the built environment relates strongly to the fields of architecture and urban planning. The built environment experiences severe impacts from disasters that have become more frequent, intense, and widespread around the world due to climate change and other human actions, requiring research on architectural design and urban planning that contributes to long-term resilience. On the other hand, the built environment itself causes environmental impacts through the consumption of energy, land-use change and pressure on ecosystems, among a variety of other factors. Two key frameworks of the United Nations—the Sendai Framework for Disaster Risk Reduction and Sustainable Development Goals—have targets concerned with the resilience and sustainability of the built environment. To meet these targets, research-informed policies, practices, and strategies need to be tailored to accommodate innovative and forward-thinking visions for a resilient and sustainable future. The built environment is multifold with its manifestation determined by economic levels, political systems, demography, and natural resources. Thus, contextual approaches and solutions in architecture and urban planning are essential, and are reflected in the diverse range of research findings consolidated in this Special Issue.

This Special Issue aims to serve as a platform to assemble recent and innovative research findings relating to resilience and sustainability in the architectural and urban planning fields, and through an open access mode, disseminate this knowledge widely at a time when the world is facing critical challenges posed by disasters and environmental transformation. The 10 papers included in this Special Issue provide insights from a range of knowledge streams and geographical contexts both from the Global South and Global North, including Australia, Israel, Italy, Nepal, Sri Lanka, USA, and Vanuatu. All the papers present issues relating to architecture and urban planning in diverse ways that have implications for resilience and sustainability. The significance of urban renewal and the related renovation of social housing have been underscored, respectively, in the papers by Shach-Pinsly and by Carrao, La Placa, Genova, and Vinci. Post-disaster recovery often provides the opportunity to build back better to enable resilience, which is evident from the improved urban planning strategies for tsunami-affected areas, as discussed by Perera, Zoysa, Abeysinghe, Haigh, Amaratunga, and Dissanayake, while the potential for self-recovery to contribute to resilient communities has been explored by Ahmed and Parrack and Carrasco and O'Brien. The importance of built environment professionals and the need for wider education in the field of resilient design have been argued for by Charlesworth and Fien. Far, Ahmed and MacKee demonstrate through their paper the significance of occupant behaviour for sustainable energy use in housing. A case study of a governmental approach to support the resilience of indigenous communities is presented by Tucker, Klerck, and Flouris. This Special Issue makes the connection to contemporary challenges such as climate change and the COVID-19 pandemic through the papers, respectively, by Pandey and Mishra and Ilatova, and Abraham and Celik.

This Special Issue presents a truly diverse and rich collection of authorships, contexts, themes and perspectives, highlighting the critical and manifold challenges facing resilience and sustainability for the built environment and communities therein, and opportunities for organisations and people worldwide to deal with these challenges are also presented. This Special Issue has relevance for the Sustainable Development Goals, which are concerned with challenges and opportunities for resilience and sustainability. Thus, this repertoire of knowledge streams resulting from multifarious research and scholarship can serve as a resource for scholars in the fields of architecture, urban planning and allied disciplines and contribute to the future discourse on resilience and sustainability that is sure to continue in a world facing escalating disaster and crisis impacts and environmental pressures.

Iftekhar Ahmed, Masa Noguchi, David O'Brien, Chris Tucker, and Mittul Vahanvati
Editors

Article

Shelter Self-Recovery: The Experience of Vanuatu

Iftekhar Ahmed [1,*] and **Charles Parrack** [2]

1   School of Architecture and Built Environment, University of Newcastle, Callaghan, NSW 2308, Australia
2   School of Architecture, Headington Campus, Oxford Brookes University, Oxford OX3 0BP, UK; cparrack@brookes.ac.uk
*   Correspondence: ifte.ahmed@newcastle.edu.au; Tel.: +61-2-4921-6011

Abstract: This paper draws from a research project that explored the lived reality of communities in Vanuatu recovering from major disasters to understand the impacts of shelter interventions by humanitarian organizations. It focuses on "shelter self-recovery", anapproach followed by organizations after recent disasters. A global overview of self-recovery highlights the potential of this approach to support recovery pathways and indicates the reliance on local context. The overview shows the need for more evidence on the impact of self-recovery programs. In Vanuatu, the study was undertaken in three island sites—Tanna, Maewo and Pentecost—affected by different disasters, particularly cyclones. It examined three main issues: (a) understanding and interpretation of self-recovery; (b) how the approach has evolved over time; and (c) what is being done by communities to support self-recovery to reduce future disaster risk. Key findings from the field indicated that devastation by disasters such as cyclones can cause a serious scarcity of natural building materials, which impedes the self-recovery process. The other significant issue is that of traditional versus modern building materials, where many people aspire for modern houses. However, poorly constructed modern houses pose a risk in disasters, and there are examples of shelters made of traditional materials that provide safety. Drawing from the field investigations, a set of recommendations were developed for more effective shelter self-recovery by humanitarian agencies in partnership with communities and other stakeholders. These recommendations place importance on contextual factors, community consultation and engagement, and addressing the supply of natural building materials.

Keywords: cyclones; self-recovery; shelter; resilience; Vanuatu

Citation: Ahmed, I.; Parrack, C. Shelter Self-Recovery: The Experience of Vanuatu. *Architecture* **2022**, 2, 434–445. https://doi.org/10.3390/architecture2020024

Academic Editor: Avi Friedman

Received: 22 March 2022
Accepted: 9 June 2022
Published: 14 June 2022

Publisher's Note: MDPI stays neutral with regard to jurisdictional claims in published maps and institutional affiliations.

1. Introduction

This paper is concerned with the process of shelter self-recovery following a disaster, which is elucidated through a case study from Vanuatu. The term "self-recovery" is used in the humanitarian shelter and settlements sector to mean the process whereby disaster-affected households repair, build, or rebuild their homes using their own resources [1,2] supplemented with assistance from humanitarian organizations. What is important about the process is that households are able to make their own choices and decisions about the reconstruction process [3]. It is an assisted self-help process, where organizations help people to help themselves, consistent with the concept suggested by Flinn et al. [4].

Self-recovery exists on a continuum from those who recover by themselves with no external input and also from programs implemented by humanitarian agencies that provide assistance to support household choice and self-determination on their path to recovery. It is not self-recovery *per se*, but an assisted self-recovery process. This support enables a self-recovery process to be delivered to a much wider percentage of the affected population than typical reconstruction programs which focus on contractor-led construction or repair of damaged houses. The self-recovery approach is related to other recovery approaches such as owner-driven reconstruction, transitional shelter, and core shelter. The main difference is that the reconstruction work aims to assist the whole of the affected population, not just those who are supported by humanitarian agencies. Even in owner-driven reconstruction projects, transitional shelter projects, or the other forms of shelter recovery projects, the focus is on humanitarian agencies supporting the construction of complete dwelling units—that is, a product—whereas the self-recovery approach is more of a process that supports disaster-affected communities and households to rebuild their own homes more safely than before through the provision of good quality tools, technical back-up, and training. In that sense, it is a different paradigm than that more widely practiced.

There is little evidence for how shelter self-recovery has been implemented and even less on measuring outcomes for self-recovery programs [3,5], and more understanding is needed. There is need for evidence of humanitarian experiences of supporting self-recovery in order to determine how self-recovery can be effective in the field, and what barriers will hamper the process. Humanitarian agencies' experience in assisting self-recovery in urban communities is limited; little is known about how to support it in practice [6]. A knowledge gap is evident: there is a plethora of literature generated in the last few decades on the operational aspects, impacts, and outcomes of shelter recovery programs consisting of projects where humanitarian agencies support the reconstruction of complete houses, but in comparison, the shelter self-recovery approach is relatively new, and there is limited research on it.

The Vanuatu case study presented here is an exploration in gathering evidence on the efficacy, challenges, and opportunities in this field, which offers lessons for the wider Global South context. Typically, as evident from the work of humanitarian organizations there such as CARE-Vanuatu (hereafter referred to as CARE), the shelter self-recovery model consists of providing emergency response support after a disaster, such as tarpaulin sheets and shelter repair kits, together with the training of local Shelter Focal Persons (SFPs) and other tradespersons, followed by the provision of basic tools and materials for house repair and reconstruction. The materials initially provided by CARE are to facilitate emergency life-saving protection from the elements, while the self-recovery process concerns the reconstruction phase of the disaster recovery cycle. The case study points to directions on shelter self-recovery for humanitarian agencies in partnership with communities and other stakeholders such as the government and private sector, articulated in this paper in the form of recommendations.

The island nation of Vanuatu, comprised of an archipelago in the Pacific Ocean, was selected as a case study because of its high exposure to natural hazards, particularly cyclones, and also earthquakes, tsunamis, and volcanic eruptions. Vanuatu is ranked as the most risk-prone country in the world [7]. Cyclone Pam in 2015 resulted in severe devastation and, while not yet fully recovered, Vanuatu was hit by Cyclone Harold in April 2020. The housing sector (or shelter) experienced massive damage—Cyclone Pam damaged 80–90% of the houses in the impacted areas, and in Cyclone Harold, more than 21,000 houses were damaged or destroyed [8,9]. This paper thus focuses on shelter in the context of such disasters.

2. Overview of Associated Literature

The justification for a self-recovery approach comes from the realization that international aid agencies direct shelter support programs that reach at most 30% of affected households in need of shelter assistance, more often with 10–20% coverage [1]. This leaves the vast majority to rebuild on their own, and thus they face significant challenges in recovery, leaving them vulnerable to future disasters, as articulated by Flinn et al. [4] (p. 12): "With little or no outside support, these families will, in most cases, rebuild their houses with the same vulnerabilities and bad building practices that had been contributory factors to the damage, economic loss, injury or death".

A study of the research priorities of expert informants [10] gives a rationale for the strong need for further research into shelter self-recovery. The expert informants identified 96 research needs which they then ranked according to a Delphi process (see for example [11] for information on this process). The issue of research into self-recovery was placed in the top three research needs. In another study, the quality of research in this area was investigated by [3], and it was found that project reports and evaluations lack sufficient detail to be able to investigate the impact of self-recovery. What is notably lacking is information on the reporting of key project activities and how outcomes are assessed.

The self-recovery approach is gaining interest from practitioners, as the research is being disseminated through practitioner networks such as the Global Shelter Cluster and the regional Shelter Forums. Maynard et al. [5] comment that shelter programs to support self-recovery are being accepted as legitimate by practitioners and gaining momentum in the shelter sector. Twigg [12] records that global institutional endorsement was given to self-recovery methods when it was adopted as a working group of the Global Shelter Cluster (the Cluster System is the operational structure of the UN Inter-Agency Standing Committee (IASC) humanitarian coordination framework).

Characteristics influencing self-recovery [2] include the recovery context, enablers and barriers, household decision making, and humanitarian support for the process. The recovery context includes access to grants, the nature of the hazard or hazards, and access to resources and community organizations. Enablers or barriers determine progress according to livelihood pressures, household priorities on how to use scarce resources, the mental ability to deal with the aftermath of a disaster, and the level of technical knowledge and skills. Decision making by households involves a complex and possibly conflicting set of dilemmas on how to navigate the recovery process. Humanitarian support for self-recovery needs to integrate different disciplines and sectors to be effective.

There are some indications of the positive aspects of agency programs that support self-recovery [4]. The training of local artisans in safer building techniques ensures that the technical knowledge for building disaster-resilient housing stays in the community and can contribute to longer-term disaster risk reduction. Each household makes decisions about where to spend the resources and design according to their specific needs, which reduces the negative effects of the imposed design of a humanitarian aid shelter program. Control over design can have the added advantage of using the shelter reconstruction to support outcomes in other sectors. An example of this process is enabling the design of the house to incorporate spaces to support livelihood activities. Speed is another positive aspect of self-recovery programming, where cash support can be delivered to large numbers of households in a short space of time. A key negative includes inconsistent technical quality. If households are recovering by themselves, they may not have the technical knowledge or the skills to be able to incorporate build-back-safer components into the reconstruction, or may not even know about safer building messages, if the dissemination of these does not reach the affected communities.

In one of the early studies of self-recovery [6], cases in the Philippines (Typhoon Haiyan response 2013) and Nepal (Gorkha earthquake response 2015) were investigated. It was found that significant barriers to self-recovery were created by local and national government policy decisions, creating an environment that did not support self-recovery. Displacing households from their livelihoods had a negative impact; it meant that although they may have had safer shelter, there was little opportunity to earn a living where they were newly located. The proposal made by [6] is that, for self-recovery to be effective, the housing need must be integrated with other needs such as including access to livelihoods, water, sanitation, hygiene, health, and food security. Cash assistance was found to have a specific negative effect on self-recovery: the most vulnerable households were not able to access the formal support systems.

In an evidence review of safety aspects of self-recovery shelter programs [3], there were only a small number of studies that reported findings in this area. The evidence was generally of poor quality, so the results need to be interpreted with caution. In [3], the authors determined three areas of relevance where self-recovery programs reported findings on the safety of the reconstructed housing: technical support, adaptation of local construction techniques, and knowledge transfer. Each of the issues are detailed below.

Technical support was very commonly used to support self-recovery through training programs, but whether this leads to safer construction seems to be related to monitoring of the construction process. Organizations supporting self-recovery programs need to have the capacity to provide technical assistance, which can be complicated if householders are given choice on which materials to use and how to fix them together.

Support for self-recovery commonly uses the adaptation of local construction techniques. The benefits of this approach are that materials are easily accessible and commonly free and can be replicated easily due to existing skills in the community, although the supply scarcity of local materials is also becoming evident, as discussed in the Vanuatu case study below. Communities are more likely to continue using these methods if they are easy to modify, so there is more chance of sustainability using these construction techniques. The main drawback of this approach is that it is time-consuming to investigate local techniques and work with the community to adapt and refine them. This time commitment may not fit easily into humanitarian reconstruction or donor financing timetables.

There is some evidence [3], although weakly presented, that indicates that knowledge transfer about safer construction can spread throughout the community to households who are self–recovering completely by themselves and do not receive any support for their recovery process. It is likely that this happens through awareness of training programs being given to households who are humanitarian program beneficiaries, or through local artisans who have received training and are then employed by households to assist with reconstruction.

An evidence review of the impact of self-recovery programs [5] similarly found only a small number of relevant and reliable reports. The review found the following ways in which shelter self-recovery can make an impact on the affected population at household level: dignity and self-reliance, perception of safety and security, income or livelihoods, assets or debts, physical and mental health, and knowledge about safer construction. The study found good support in the evidence for positive effects of self-recovery programs on dignity and self-reliance due to people taking ownership of the rebuilding process. It also found good support for the perception of better safety and security. This was as a result of reduced overcrowding, integration into host communities, and awareness of what constitutes reliable materials and good construction quality. The evidence for the other factors was inconsistent or unclear and needs more research to explore.

3. Materials and Methods

Informed by a broader study on self-recovery supported by the Global Challenges Research Fund (GCRF) on "Self-Recovery from Humanitarian Crisis" conducted by CARE-UK and Oxford Brookes University, UK [13], of which the Vanuatu case study below is a part, and in response to the need for evidence mentioned in the previous section, the following three research questions were investigated in the field studies in Vanuatu and the analysis of the data collected therein:

1. How is self-recovery understood by different stakeholders? (RQ1)
2. How has community self-recovery and support for self-recovery evolved? (RQ2)
3. What interventions and strategies are used by communities to support shelter self-recovery and enhance disaster risk reduction (DRR) practice? (RQ3)

The basis of RQ1 is linked to the limited existing research on self-recovery, as discussed above in Section 2, and thus in Vanuatu, it was sought to understand the perspective of both community and organizational representatives. There are reports that the self-recovery approach had been applied in recent years in Vanuatu, especially since TC Pam in 2015; given that TC Harold was a recent disaster, RQ2 was posed to explore the most recent developments in self-recovery. While self-recovery is essentially an agency-assisted process, the community plays a vital role in it; therefore, RQ1 explored that role and at the same time explored DRR practice within the self-recovery paradigm.

The research was conducted in partnership with CARE and mainly explored the self-recovery initiatives of this organization; the work of some other organizations was also investigated. Investigations were undertaken on three different islands in Vanuatu—Tanna, Maewo and Pentecost—that were affected by different disasters and where self-recovery interventions were implemented. Tanna was affected by TC Pam in 2015, and the work of CARE there represents an earlier stage of the application of the self-recovery approach. In Maewo, people were relocated from the nearby island of Ambae after volcanic activity from 2017, and the self-recovery work by CARE consisted of support for the new settlers to build houses that would provide safety from future disasters, which the existing community had the opportunity to utilize as well. Pentecost was severely affected by TC Harold in 2020, and CARE undertook an extensive shelter self-recovery initiative there that included a variety of inputs including the provision of training, materials, and equipment.

The data collection methods were adapted from methodological approaches developed by the GCRF team. The communities participating in the research were purposely selected based on their involvement in CARE's humanitarian response programs. A representative sample of research participants included a range of responses and experiences, a gender balance, and a diversity of geographical locations (e.g., inland and coastal communities, north and south situated communities). The data collection methods included focus group discussions (FGDs) at the community level, semi-structured interviews of households and community key informants on the three islands, and semi-structured interviews of organizational key informants in Port Vila, Tanna, and overseas (see Table 1). The interview and FGD questions were structured along the lines of the three research questions above: firstly, to find out about the understanding of shelter self-recovery of the participants at both community and organizational levels; secondly, to know about their respective experiences in the series of recent disasters (Cyclone Pam, Cyclone Harold, Ambae Volcano) with a view to explore the evolution of shelter self-recovery; and thirdly, focusing on local support strategies for shelter self-recovery with a view to understanding DRR measures. A comparison of the responses by the two different groups—communities and organizations—allowed a comparison of the different and/or similar perceptions and interpretations, thereby contributing to the research results.

Thematic analysis was conducted on the collected data, which is suited to this from of phenomenographical research, which sought to examine how different people understood and experienced the phenomenon of shelter self-recovery (see [14] on this form of research). In practical terms, it involved a manual coding method to identify key themes. The thematic codes consisted mainly of perspectives of the respondents, and thereby recurrent and also unique issues were identified. The interview and FGD responses were reviewed to identify a set of recurrent issues pertaining to shelter self-recovery in Vanuatu, which were manually assigned individual color codes, which then informed the structure and components of the research results. Unique issues were also identified to complement the narratives from the common issues.

A qualitative approach was followed, involving both secondary and primary data collection and the analysis of the data to identify key issues relating to shelter self-recovery. The case study presented below is derived mainly from the primary data from the field, consisting of interviews and focus group discussions (FGDs), and also informed by some key relevant documents.

Table 1. Summary of data collection methods.

		i. Community Level		
	Methods	**Pentecost** **6 Communities (FGDs** **in 5)**	**Tanna** **2 Communities**	**Maewo** **3 Communities**
1.	**Focus Group Discussions (FGDs)** (total 20 FGDs in 3 locations)	1 female + 1 male FGD per community. TOTAL 10 FGDs	1 female + 1 male FGD per community. TOTAL 4 FGDs	1 female + 1 male FGD per community. TOTAL 6 FGDs
2.	**Semi-structured interviews:** Community Key Informants, especially Shelter Focal Points (SFPs) (total 15 interviews in 3 locations)	7 interviews (3 female)	2 interviews (1 female)	6 interviews (1 female)
3.	**Semi-structured interviews:** Households, including vulnerable people such as the elderly, widows and people with disabilities (total 10 interviews in 3 locations)	6 interviews (4 female)	2 interviews (1 female)	2 interview (2 female)
		ii. Organization Level		
	Method: Semi-Structured Key Informant Interviews (KIIs) (TOTAL 10 interviews)			
1.	National Disaster Management Office	1 × Staff (Port Vila)		
2.	Department of Strategic Planning and Aid Policy Coordination (DESPAC)	1 × Staff (Port Vila)		
3.	Public Works Department/Shelter Cluster Lead	1 × Staff (Port Vila)		
4.	Lume Rural Training Centre (LRTC)	1 × Staff (Tanna)		
5.	International NGO (CARE)	2 × Staff (Port Vila)		
6.	Vanuatu Red Cross Society (VRCS)	1 × VRCS Disaster Manager with Shelter/PASSA/field experience (Port Vila)		
7.	International NGO (IOM)	1 × Staff (Port Vila)		
8.	International NGO (Australian Red Cross)	1 × Staff (Australia)		
9.	Independent Shelter Consultant	1 × Consultant (Port Vila); former CARE Shelter Technical Adviser		
10.	Donor agency	1 × USAID (Washington D.C.)		

4. Results

The results of the research are structured below according to the three research questions mentioned above in Section 3. These results were the basis for identifying the recommendations presented later in Section 5. Some of the key issues that emerged from the field investigations, as discussed below, underscored the importance of understanding the context, community consultation and engagement, and the supply of natural building materials, which have informed the formulation of the recommendations.

4.1. Understanding Self-Recovery

The initial community recovery process evident from Vanuatu consists of salvaging materials from damaged homes to build a temporary shelter. People might live in inadequate shelter over an extended period after a disaster as they incrementally improve their dwellings. The self-recovery projects of organizations such as CARE focus on incorporating a degree of safety so that these houses can be resilient. Tarpaulin sheets provided as emergency relief assist in protection in the interim (see Figure 1) before more durable options are available through the support of agencies including training and tools. The shelter self-recovery model generally does not directly provide building materials or cash, and disaster-affected households are expected to source the materials themselves. Tarpaulin sheets are provided as emergency relief for temporary shelter during the crisis stage, although households have been found to be using them well after the initial period.

Figure 1. A temporary house built with salvaged materials after TC Harold in Pentecost. The structure on the left has tarpaulin sheets provided as emergency relief by CARE (photo credit: Jen Bowtell).

There appeared to be a lack of clarity at the community level on the concept of self-recovery as an assisted self-help process, indicating the need to translate it to that level. At the level of organizational staff, more clarity was evident, and several staff members highlighted the merits of the approach.

The issue of traditional versus modern building materials is significant in this context, where many people aspire for modern houses. However, poorly constructed modern houses pose a risk in disasters, and there are examples of shelters made of traditional materials that provide safety. Affordability is a key issue with low cash flows in rural areas; thus, humanitarian organizations such as CARE aim to support the use of traditional materials with safety features using a limited set of modern materials such as cyclone straps, wire, and nails, together with training based on Shelter Cluster Vanuatu (SCV—a consortium of humanitarian organizations involved in the shelter sector) guidelines on safe construction. This, however, causes tension between local aspirations and the aims of humanitarian agencies; thus, again, the significance and necessity of community consultation and engagement is highlighted in the recommendations below.

4.2. Evolution of Self-Recovery

A key purpose of shelter self-recovery is to enable communities to build back safer (BBS), which has stepped up in Vanuatu since TC Pam. This involves incorporating safety features in shelter that allow a degree of resistance and reduce the need for repair and rebuilding after a disaster. The BBS message was found to have gained ground through CARE's efforts in the research sites.

The supply of natural materials, as used in traditional houses in Vanuatu (see Figure 2), is important for self-recovery but can be affected by the loss of vegetation in a disaster; people in Pentecost were facing an acute shortage of *natangora* (sago palm) thatching material. CARE trained and supported chainsaw operators to harvest timber from trees fallen by the cyclone, which addressed the problem of materials shortage, though people needed to wait for some years before the *natangora* and bamboo groves grew back and could be used for further rebuilding. There is an overall growing problem of a diminishing supply base of natural materials, pointing to the need for sustainable forestry initiatives.

(a)   (b)

Figure 2. Traditional houses made of natural building materials in (**a**) Pentecost and (**b**) Tanna (photo credits: Jen Bowtell).

A review of organizational interventions indicated gaps in the government's work at different levels. While there were critiques of the work of NGOs, there was also appreciation. The SFP model that evolved since TC Pam was found to be valuable, as well as the overall self-recovery support by CARE, because of the proximity and access to the SFPs. However, it was found that organizational support can create long-term expectations that might be unrealistic, so a clearly defined exit strategy for specific self-recovery programs that is understood and accepted by the community is important.

4.3. Knowledge Exchange and Shelter DRR

An important element in shelter self-recovery is knowledge exchange to develop local capacity and awareness in safe building practices. This is undertaken by organizations such as CARE through training programs and demonstration by building safe houses. The training of SFPs is important in this regard, as well as others such as chainsaw operators. Strong resilience and traditional knowledge exist in communities, which is a resource that self-recovery initiatives draw upon.

The role of different communities is illustrated by the specific case of relocation in Maewo. Here, the original plan of the government to establish a large permanent settlement did not materialize, and people were disappointed by the failed promise. This impacted a self-recovery project of the government where people could not or did not complete houses for which a structural frame was provided. The relocation created pressure on the existing community, but even then, there was evidence of strong community-to-community support, where the host community extended a wide range of supports to the relocated community, such as food, accommodation, and employment.

The gender strategy of CARE had resulted in positive outcomes in a strongly male-dominated context. This is particularly reflected in the development of a cadre of trained women SFPs, which was recognized by the local male Chiefs. There were still challenges with the continuing economic reliance on men and limited involvement of women in machinery-related trades, pointing to the need to continue gender-based initiatives.

Low income is a key factor contributing to vulnerability, which is compounded by various factors such as being a woman, elderly, single mother and/or widow, or having a disability and limited access to knowledge and skills. The self-recovery projects of organizations such as CARE therefore place priority on addressing the needs of the most vulnerable. At the same time, there was strong social capital in communities ranging from the family to community to community-to-community levels—a valuable resource for self-recovery programs.

Incorporating DRR features into shelter is a key aim of self-recovery programs. In addition to physical improvements to shelter, training at the community level based on safe construction guidelines had raised community awareness. This was augmented by traditional knowledge. Traditional safe refuges are generally sturdy and can serve as evacuation centers during a cyclone, and building demonstration safe houses by CARE in places such as Maewo, where they were not common, created further awareness of shelter DRR (see Figure 3).

Figure 3. An improved version of a traditional house in Maewo that was built after training by CARE (photo credit: Jen Bowtell).

5. Recommendations

Following from the preceding sections, in this section, a set of recommendations for humanitarian organizations working in the shelter sector informed by the results of the research are presented. The recommendations are drawn, interpreted, and augmented from the commentary of research participants at different levels—community and organizational—giving them a voice in contributing towards defining the future pathway of shelter self-recovery.

- Understanding the context: Every shelter self-recovery program is different, and there is a need to predict what will work and what will not according to the context. There is a need to think about the context, about what the community wants, what they are used to, and what materials and labor (skilled or unskilled) are available.
- Consulting and communicating with the community: Consulting and communicating with the community for program delivery is widely understood in the NGO community; yet, misunderstandings can occur because of differing expectations, which need to be managed to avoid negative outcomes. The idea of self-recovery needs to be translated at the community level so that there is stronger clarity and undue expectations are not created.
- Drawing on community resourcefulness: Social capital, traditional knowledge, and community leadership are key areas of community resourcefulness that inform shelter self-recovery initiatives. The self-recovery model should rely on a synthesis of organizational support with community resourcefulness.
- Considering knowledge exchange needs: A community consultation process allows insight to be gained into the knowledge exchange needs at the community level, as evident from the field investigations, where the need for refresher trainings, certification of SFPs, guidance on different types of construction—both modern and traditional—were some of the suggestions that deserve consideration. A key area to explore is additional knowledge on how to source materials beyond an island through community–community support.

- Promoting the necessity of disaster preparedness: With regard to shelter self-recovery, specific measures for disaster preparedness need to be promoted at the community level, where people require more training and awareness to ensure appropriate actions relating to shelter, food gardens, water supply, etc. in a disaster, and preparation of adequate safe shelters and other concerns in a disaster. Preparedness for a supply of natural materials after a disaster is also important by exploring options for support to local suppliers.
- Supporting sustainable forestry and plantations: To address the materials supply crisis, natural resource management including replanting and rejuvenating trees for timber supply and also natural organic materials—bamboo, *natangora*, wild cane, plants for bush rope, etc.—according to local building traditions of specific island contexts is necessary. This can be done in partnership with the government.
- Advocating for governmental engagement: NGOs should be advising and helping the government to set policies and help people access skills, materials, and logistics that support self-recovery strategies and provide guidance on means of leveraging financial and institutional resources. Advocating governmental support to Community Disaster and Climate Change Committees (CDCCCs) under the National Disaster Management Office (NDMO) to engage in self-recovery is important. Other areas include advocacy to include traditional shelter building techniques and materials in building codes and education.
- Continuing support for gender and social inclusion initiatives: Humanitarian organizations such as CARE should build on their achievements in the field of gender and social inclusion and use these as a tool for wider leverage at the institutional and community levels. They need to continue and develop this role so that women, the elderly, and people with disabilities have a stronger voice in shelter self-recovery.
- Assessing the applicability of a cash transfer approach: Although the self-recovery model followed by organizations such as CARE does not provide cash, a study by Oxfam [15] suggests that cash transfers can be feasible in some cases. It thus needs to be assessed whether a cash transfer approach is applicable according to the context. It might be applicable especially for the most vulnerable, and it also allows making individual choices on how to recover according to specific circumstances. Conditional cash transfer can be tied to the application of resilience measures in shelter. This might mean looking further at training and access to knowledge for a self-recovery approach and options for a recovery grant process.
- Having an exit strategy with strong legacy outcomes: It is important to plan an exit strategy after project completion with valuable legacy outcomes. What is left behind needs to be taken into account, and the level of self-reliance and empowerment for future self-recovery projects needs to be understood. The possibility of future ongoing shelter maintenance through support by SFPs should be explored.

6. Conclusions

This paper provides much-needed evidence on the application of a self-recovery approach in post-disaster shelter assistance. The literature highlights the need for better quality evidence, and this study demonstrates a rigorous scientific method and analytical framework. The research approach highlights the strength of research collaboration between operational humanitarian agencies and academic organizations, harnessing the strengths of both to enable quality research. The connections and country office of CARE in Vanuatu enabled dependable groups of individuals and communities to be identified and accessed through their network of beneficiaries. The academic partners enabled the research method to be set up and analyzed in a rigorous manner to ensure good-quality evidence was produced. An additional benefit is the research impact. An operational partner organization who has been operating in the area for a long time period will continue to be present in the future. The findings and lessons from the research can therefore be directly applied to planning and response to future disaster events and incorporated into policies and standard operating procedures of the partner organization.

From this case study, the understanding of self-recovery has been advanced. In Vanuatu, it was found that households aspired to live in houses built from "modern" materials using "modern" construction methods, which caused tension with the proposals of the agency supporting the self-recovery process, which focused on traditional materials and construction methods with added safety features. It is likely that this issue exists more widely that Vanuatu, but more research would be useful to determine this, to address what can be done to manage expectations during the recovery process, and to communicate the value (sustainability, maintenance, cost) of the safer traditional construction approach.

There is a key relationship between self-recovery and the natural environment evidenced in this research. The self-recovery approach, by prioritizing traditional materials obtained from the natural environment, highlighted future shortages of these materials unless policies and projects were put in place to ensure the sustainable supply of these materials. One positive aspect of this linkage is that livelihood programs to work with the natural environment to produce building materials could be a positive outcome of support for self-recovery housing. Again, this is an issue that has the potential to be significant for the protection of environmental resources as well as life cycle environmental impacts of housing programs in many situations of housing recovery. More research is needed in this area to determine the extent of this relationship and measure its impact.

Finally, the importance of knowledge exchange is highlighted in the research findings. Self-recovery depends on information being spread amongst the affected population on better and safer reconstruction practice, so this represents a key area of investigation. In the case of Vanuatu, the Shelter Focal Person (SFP) role in the community represents an opportunity to locate knowledge in the structure of the community, and the creation of this role ensures continuity. Successful knowledge exchange was also seen in community-to-community support and the knowledge exchange between safer construction guidelines and traditional construction knowledge. All of these factors be worthy of study in other geographical contexts.

Author Contributions: This paper is derived from a consultancy report produced by I.A., who was responsible for conceiving the paper and writing the introduction, methodology, results, and recommendations sections. C.P. wrote the literature review and conclusions section and also provided some inputs to the other sections. All authors have read and agreed to the published version of the manuscript.

Funding: Funding for this research was provided by the UKRI Global Challenges Research Fund, Global Research Translations Award. Reference EP/T015160/1.

Institutional Review Board Statement: Research ethics approval for this project has been awarded by Oxford Brookes University Ethics Committee, UREC Registration No: 201465.

Informed Consent Statement: Research ethics protocols for informed consent were followed as per the above-mentioned ethics approval from Oxford Brookes University, UK.

Data Availability Statement: The full report and related information is available from the "Self-Recovery" website of Oxford Brookes University: https://self-recovery.org/ (accessed on 8 June 2022).

Acknowledgments: Acknowledgements are due to CARE-UK and CARE-Vanuatu teams for supporting the research project in Vanuatu and for their feedback during the process of producing the project report. Lead persons include Bill Flinn (CARE-UK) and Isabelle Choutet (CARE-Vanuatu), who played key roles in supporting and facilitating the project, as well as providing valuable conceptual feedback to the study. Acknowledgement also goes to Jen Bowtell and her team in Natapoa Consulting, Vanuatu, who was the Research Associate and conducted the data collection.

Conflicts of Interest: The authors declare no conflict of interest.

References

1. Parrack, C.; Flinn, B.; Passey, M. Getting the message across for safer self-recovery. *Open House Int.* **2014**, *39*, 47–58. [CrossRef]
2. Twigg, J.; Lovell, E.; Schofield, H.; Miranda, M.L.; Flinn, B.; Sargeant, S.; Finlayson, A.; Dijkstra, T.; Stephenson, V.; Albuerne, A.; et al. *Self-Recovery from Disasters: An Interdisciplinary Perspective (Working Paper 523)*; Overseas Development Institute: London, UK, 2017.
3. Harriss, L.; Parrack, C.; Jordan, Z. Building safety in humanitarian programmes that support post—Disaster shelter self-recovery: An evidence review. *Disasters* **2020**, *44*, 307–335. [CrossRef] [PubMed]
4. Flinn, B.; Schofield, H.; Morel, L.M. The case for self-recovery. *Forc. Mig. Rev.* **2017**, *55*, 12–14.
5. Maynard, V.; Parker, E.; Twigg, J. *The Effectiveness and Efficiency of Interventions Supporting Shelter Self-Recovery Following Humanitarian Crises*; Oxfam: Oxford, UK, 2017.
6. Schofield, H.; Lovell, E.; Flinn, B.; Twigg, J. Barriers to urban shelter self-recovery in Philippines and Nepal: Lessons for humanitarian policy and practice. *J. Brit. Acad.* **2019**, *7*, 83–107.
7. Radtke, K. *World Risk Report*; Ruhr University Bochum: Bochum, Germany, 2019.
8. Shelter Cluster. Shelter and Settlements Vulnerability Assessment—Final Report: Cyclone Pam Response. 2015. Available online: https://www.sheltercluster.org/sites/default/files/docs/shelter_cluster_report_shelter_and_settlements_vulnerability_assessment_after_cyclone_pam_may2015.pdf (accessed on 7 April 2022).
9. Shelter Cluster. TC Harold Sitrep. In Proceedings of the Academic-Practitioner Collaboration for Urban Settlements—South Pacific (APCUS-SP), Online, 12–21 June 2020.
10. Opdyke, A.; Goldwyn, B.; Javernick-Will, A. Defining a humanitarian shelter and settlements research agenda. *Int. J. Disaster Risk Reduct.* **2021**, *52*, 101950. [CrossRef]
11. Twin, A. Delphi Method. Available online: https://www.investopedia.com/terms/d/delphi-method.asp (accessed on 17 March 2022).
12. Twigg, J. The evolution of shelter "self-recovery": Adapting thinking and practice for post-disaster resilience. *J. Brit. Acad.* **2021**, *9*, 5–22. [CrossRef]
13. Oxford Brookes University. Self-Recovery. Available online: https://self-recovery.org/ (accessed on 17 March 2022).
14. Richardson, J.T.E. The concepts and methods of phenomenographic research. *Rev. Educ. Res.* **1999**, *69*, 53–82. [CrossRef]
15. Holt, C.; Hart, S. *Vanuatu Cash Transfer Feasibility Assessment*; Oxfam Vanuatu: Port Vila, Vanuatu, 2019.

Article

Incremental Pathways of Post-Disaster Housing Self-Recovery in Villa Verde, Chile

Sandra Carrasco [1,*] and David O'Brien [2]

1   School of Architecture and Built Environment, The University of Newcastle, Newcastle, NSW 2308, Australia
2   Faculty of Architecture, Building and Planning, The University of Melbourne, Parkville, VIC 3010, Australia; djobrien@unimelb.edu.au
*   Correspondence: sandra.carrasco@newcastle.edu.au or sandramcarrasco1@gmail.com

Abstract: Housing reconstruction is considered the backbone of disaster recovery. The increasing losses in housing due to disasters challenge conventional top-down schemes and call for people-centred approaches to acknowledge their agency and self-recovery resources. This paper examines the pathways for housing self-recovery through resident-controlled incremental housing development. This paper focuses on the Villa Verde settlement built in the Chilean city of Constitución, which was severely impacted by the 2010 Chile Earthquake. Villa Verde, designed by the Chilean architecture studio Elemental, is one of the most notable incremental housing projects worldwide that encourage residents to extend their houses within a provided structural framework. This research aims to provide clarity in the much-needed understanding of disaster-affected people's agency to self-recover, noted by researchers as one of the crucial elements for improving the humanitarian response in the aftermath of disasters. Through capturing the evolution of incremental housing construction, this paper presents multiple complexities resulting from the variety of households' characteristics and needs in their process of post-disaster housing self-recovery. The resident-controlled process studied evidence that the people's capacities and dedication to self-recover challenged the established housing framework with extensions beyond the designers' parameters requiring further evaluation of the long-term implications of self-help constructions.

Keywords: Elemental; Villa Verde; incremental housing; self-recovery; adaptation

Citation: Carrasco, S.; O'Brien, D. Incremental Pathways of Post-Disaster Housing Self-Recovery in Villa Verde, Chile. *Architecture* **2022**, *2*, 544–561. https://doi.org/10.3390/architecture2030030

Academic Editor: Miguel Amado

Received: 12 July 2022
Accepted: 28 July 2022
Published: 29 July 2022

Publisher's Note: MDPI stays neutral with regard to jurisdictional claims in published maps and institutional affiliations.

1. Introduction

Disasters have tremendous impacts on housing. Housing accounts for up to 62% of economic losses due to disasters [1]. Global estimations indicate that more than 5 million houses were destroyed or collapsed due to natural hazards and conflicts between 2005 and 2018 [2]. Predictions of future disaster-related housing losses indicate that up to 167 million houses will be destroyed in the next two decades [3]. Housing damages are expected to be magnified by their unpredictable frequency and intensity, especially climate change-related events. Besides the significant economic impacts, housing loss due to disasters has profound and cascading effects on affected communities' dynamics [4] and generates traumatic changes in survivors' lifestyles and living conditions [5] that can take many years to recover. Therefore, re-establishing housing is considered the backbone of recovery impacting different dimensions of individual, household and community recovery [6,7]. Moreover, it cannot be ignored that for disaster-affected people the loss of their homes represents much more than the loss of a physical structure. Housing embodies the place where people carry out their daily activities, domestic functions, routines, homes also become workplaces and sources of income [6,8,9].

Despite the urgency in the timely provision of adequate housing following a disaster, studies have acknowledged the limitations of the humanitarian sector and government agencies [10]. The top-down constructor-driven reconstruction is one of the most criticised but predominant approaches in post-disaster housing reconstruction. The limited or no space for affected communities' engagement and failure to acknowledge the diversity of people's needs and priorities questioned the top-down approach capacities to support recovery and build resilient communities [11,12]. Despite the discussions about the most appropriate housing reconstruction schemes, the reality of post-disaster housing recovery reveals that most of the affected people rebuild their houses using their own resources and capacities [13]. Recent estimations suggest that between 8 and 9 in 10 destroyed or damaged houses are repaired or rebuilt by the affected families with little or no outside support through 'self-recovery' [10].

Observing the characteristics of self-recovery, based on people's initiatives which might include or not external support from humanitarian agencies or governments [14], it is unavoidable to see how self-recovery emulates the resident-controlled incremental housing process, although in a post-disaster context. Incremental housing is a people's centred and controlled process of housing production responding to the inadequate and insufficient housing stock in the Global South, where it is estimated that 50–80% of people build incrementally [15]. According to Nohn and Goethert [16], incremental construction is the largest viable creator of new affordable housing as residents use the "pay as you go" principle to gradually build and improve their homes based on available capacities and resources at the time. Interestingly, Latin America is considered the "birthplace of ideas for legitimising self-help housing approaches" [17] through government policies promoting owner-driven housing development by providing "site and services", expandable "core houses" or subsequent subsidies for housing improvement or extensions [17] influenced by the seminal studies conducted by John Turner [18]. Different types of incremental housing schemes have been implemented in Latin America since the 1950s, and incremental housing programmes have been promoted in Chile since the 1990s [19]. The Chilean Pritzker-awarded architectural studio Elemental is one of the advocates of incremental housing, revisiting the idea of partially completed housing inspired by John Turner's claims that people should have the "freedom to build" [20]. Elemental proposed a structural framework for residents to gradually build extensions based on the Dutch architect John Habraken's proposal in his book "Supports" [21] to facilitate a support structure to be infilled by residents who control their construction. The Elemental-designed project of Quinta Monroy in 2005 became an iconic example of in situ replacement of informal housing within an incremental framework. Elemental's approach became a reference for social and participative housing discussion in Chile and internationally [22–24].

The 2010 Chile earthquake and tsunami devastated various coastal cities in central Chile. The city of Constitución was one of the most affected, with almost two-thirds of its housing stock damaged or destroyed [25]. In the aftermath of the 2010 Chile Earthquake, reconstruction plans for the city of Constitución focused on incremental housing schemes influenced by Elemental's partially built housing framework promoting active engagement of disaster-affected communities in the co-production of their homes [26]. The Villa Verde settlement developed in a government-corporation partnership was developed as a social housing project to resettle disaster-affected and social housing beneficiaries. The housing reconstruction projects in Constitución prioritised conventional top-down approaches. Conversely, Villa Verde embraced the lessons from Chilean incremental housing projects and the structural framework to be infilled by residents based on their personal needs and aspirations for post-disaster housing.

Developing appropriate post-disaster housing approaches and their long-term impacts on affected communities are some of the main concerns of humanitarian agencies, governments and stakeholders involved in post-disaster reconstruction. Experts point out the intricacies of the evolution of self-recovery and call for a longer-term perspective on the priorities for learning and enabling change in focus toward safe, durable and sustainable recovery [27].

This research aims to clarify the much-needed understanding of disaster-affected people's agency to self-recover, noted by researchers as one of the crucial elements for improving the humanitarian response in the aftermath of disasters [12–14,27]. This research examines the pathways for shelter self-recovery through resident-controlled construction of post-disaster housing within Elemental's half-built incremental housing framework, including disaster-affected and unaffected households relocated in Villa Verde. By analysing the temporality of the housing extensions, this study observes the impact of changes in household makeup and financial conditions on the physical changes of the houses.

2. Research Design

2.1. Sample and Participants' Selection

The data for this paper was collected between July and August 2017. The complexities in understanding the process of incremental housing construction in Villa Verde require mixed-methods research combining a self-administered questionnaire survey and face-to-face semi-structured interviews. Participants selected included disaster-affected and non-affected families, as both were eligible to be beneficiaries of a house in Villa Verde. However, many interviewed families also expressed that they applied on both grounds.

The initial contact with the residents of Villa Verde was with the community leaders. At the moment of the fieldwork, there were two neighbours' associations. One of these associations' leaders requested help from us to design an extension of the communal centre after they knew our architectural background. We understood this as the first sign of residents' motivation to improve their neighbourhood and their houses. Later, during the interviews, we received some individual requests from the residents to help with their houses' extensions and renovations design.

Community leaders introduced some participants individually and during a community meeting with residents in middle July 2017. Other participants were randomly selected based on their availability. In total, 40 residents agreed to undertake the questionnaire and among them 31 residents agreed to be interviewed. The detailed contents of the tools used for data collection are described below:

(1) A closed question quantitative questionnaire seeking (a) general household information, (b) previous housing conditions, (c) initial housing conditions after moving to Villa Verde, and (d) current housing conditions.

(2) A qualitative semi-structured interview aiming to expand the knowledge of housing modifications not captured in the questionnaire. The interview included (a) timelines for incremental additions, (b) difficulties encountered when undertaking incremental improvements, (c) plans for future improvements, and (d) changes to the broader neighbourhood. This information was graphically captured through diagrams based on the methodology by the Special Interest Group in Urban Settlement, Massachusetts Institute of Technology [28–30].

(3) Mapping of settlement conditions using architectural research tactics, including architectural drawings, photographic surveying, and physical trace analysis [31–33].

(4) Secondary data from government reports, books, and academic papers.

2.2. Data Analysis

The information analysis is based on the premise that the complexities and multidimensional approach needed to analyse the interactions between people and the built environment require diverse research strategies [34,35]. This paper primarily considers the qualitative component analysed at household level using data collected through (2) and (3) specified in the previous section and uses the quantitative component as supporting data at settlement level.

This paper analyses ten families (see Figure 1) and includes their narratives related to their housing modification over time. The reason for combining the narratives with four-year timelines overlapping changes in household size, financial situation and housing extensions/modifications responds to two concerns:

(1) The spontaneous nature of the incremental construction can only be captured by understanding the specific circumstances and context in which the changes that otherwise would be misrepresented or omitted in quantitative studies [36].

(2) The clear patterns observed in these ten selected households provide a realistic image of the process of incremental housing at the household level.

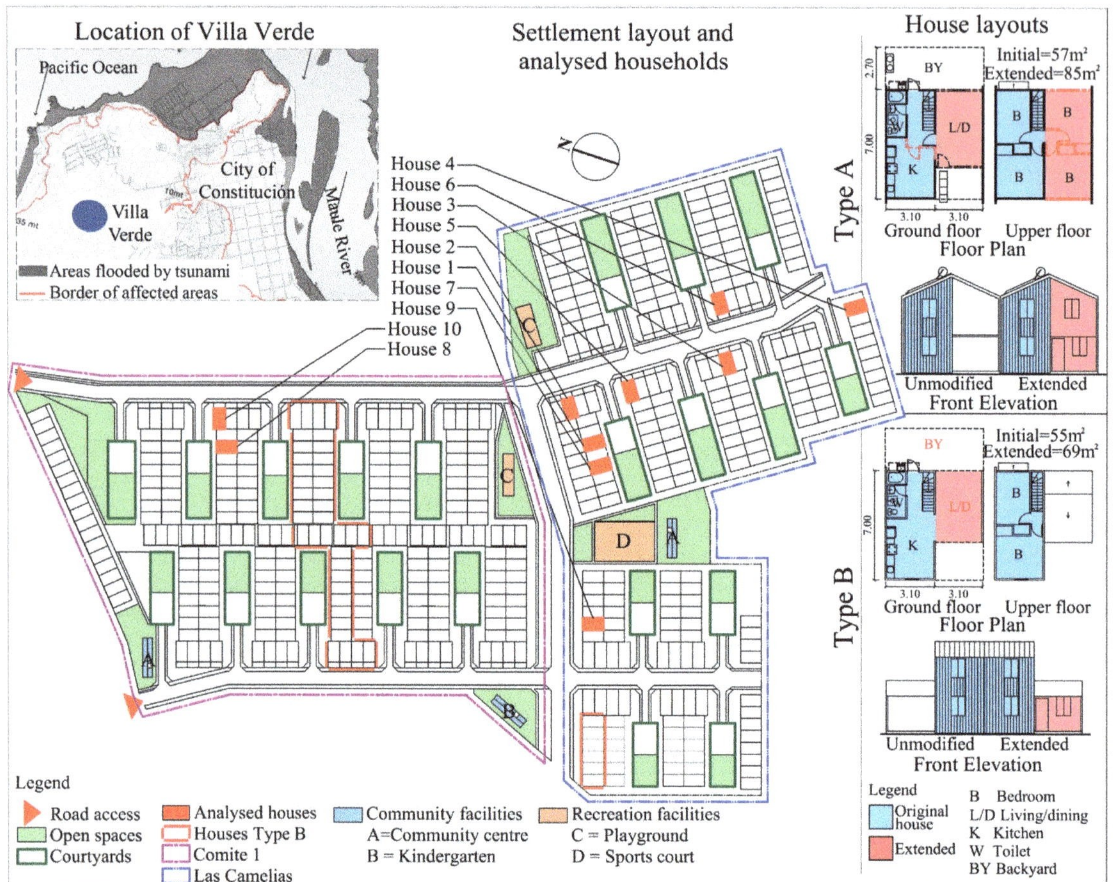

Figure 1. Settlement and housing layout of Villa Verde. Source: Authors, based on Aravena & Iacobelli [37].

2.3. Scope and Limitations of the Study

This paper focuses on a specific project and covers a small sample size relevant to understanding household conditions changes. The scale of the analysis is limited to each household and the changes made in their housing. Therefore, this paper does not attempt to generalise the challenges faced by all the residents in Villa Verde or other housing projects designed by Elemental.

3. The Housing Project at Villa Verde

3.1. Context and Planning

The housing complex in Villa Verde is part of the "Plan for Arauco Forestry Workers" promoted by the forestry company Arauco as an initiative to provide definitive houses for their workers in Chile's southern region Maule in 2009. Villa Verde was conceived as a private-public funded housing project in 2009, and Elemental was commissioned to develop an incremental housing project [37].

The earthquake and tsunami that hit the southern regions of Chile in 2010 changed the project's context. The city of Constitución was one of the most affected cities by the disaster, where the tsunami flooded 61% of the city's urban area [25]. Constitución's reconstruction and the allocation of disaster-affected families in permanent housing became the priority of the post-disaster recovery planning managed with a public-private partnership responsible for delivering the Master Plan for Sustainable Reconstruction (PRES Constitución—in Spanish) involving Government Agencies, the company Arauco and international consultants including ARUP and Elemental. The PRES Constitución presented Elemental's incremental housing designs to replace the city's destroyed housing stock [38]. However, only two incremental housing projects designed by Elemental were built in the city, Villa Verde, and its second stage under construction in 2017. The disaster changed the beneficiaries of the houses in Villa Verde, which combined Arauco workers and disaster-affected people, for this reason Villa Verde is considered a post-disaster housing and received further government subsidies for its construction through the Housing Solidarity Fund programme (FSV—in Spanish).

The Housing Solidarity Fund targets the most vulnerable first and two quintiles, meaning the country's 40% poorest. Under the Housing Solidarity Fund conditions, housing beneficiaries must co-fund the housing projects [39]. In Villa Verde, beneficiaries made a single payment of 400,000 Chilean Pesos, approximately 590 USD, and received their houses' formal ownership although subjected to some conditions. Subsidies' conditions prevented beneficiaries from renting the houses out within five years and selling the property in the following ten years from the turnover date.

Beneficiaries are also requested to follow the Chilean regulations for a joint property since Villa Verde is a residential condominium essential for constructing extensions and housing modifications. Elemental's claimed that they prioritised community participation through the beneficiaries' involvement in the planning stage of Villa Verde. The consultations' outcomes with the beneficiaries were the housing designs, including the specific housing extensions allowed to be built by themselves [37]. In the turnover ceremony, each household received a "Habitability Manual" that included directions for housing maintenance and the specifications of the extensions approved by the city council in the housing building permit. As Villa Verde falls under a condominium property scheme and following the Chilean regulations, housing modifications not specified in the building permit required the neighbours' collective endorsement and formalisation following the local city council procedures [40]. The Housing Solidarity Fund program also allows for further government subsidies such as the Housing and Neighbourhood improvement program [41], although it is limited to works within the formally approved designs. Furthermore, the Housing and Neighbourhood improvement program specifies that only collective applications are received for housing improvements [41]. The regulation limits applications to neighbours' associations or community associations established within the settlement, for instance, in Villa Verde, there is a seniors' neighbours association that exists in parallel to the two neighbours' associations.

3.2. Design

Villa Verde is located in a high hill area approximately 2.5 km from the city centre and next to consolidated residential subdivisions (Figure 1). Villa Verde is made up of rows of 484 two-story houses organised in clusters sharing one common courtyard. There are two typologies of houses, Types A and B (87% and 13%, respectively) (Figure 1). Elemental's designs show that the Type A half-house was built to 57 m^2 and is possibly expanded to 85 m^2. The 62 Type B houses, initially built at 55 m^2, are expandable to 69 m^2. For this study, only house Type A is considered.

The housing design (Type A) considers a wooden structure covered by a gable roof in which half of the house is infilled and the other half is left empty for the residents to build extensions (Figure 1). A kitchen, small dining area, and a bathroom were built on the lower floor while the living room was expected to be built by the residents. On the upper floor two bedrooms were built and other two were considered for residents to extend. Prefabricated panels covered the houses' wooden structure as the finishing of walls, cement floors, and lacking finishing. Kitchens came with just a sink without cabinets or working tables. The Habitability Manual discourages residents from building in the backyard or in the frontal garden arguing that habitability conditions such as natural lighting and ventilation would be compromised.

3.3. Beneficiaries' Selection and Housing Allocation

Following the 2010 disaster, 244 disaster-affected families were included as beneficiaries alongside 240 Arauco households and in many cases some households qualified on both grounds. Beneficiaries had the opportunity to show a preference for the type of house and its location in the settlement. Priority was given to beneficiaries based on their contribution during community meetings and workshops with Elemental and Arauco.

4. Incrementality and Resilience in Villa Verde

The timeframes and continuity of incremental housing production are not homogeneous and respond to the changing conditions and needs resulting from the process of human habitation [42,43]. Furthermore, in Villa Verde, homeowners had three years between house planning and turnover to prepare for building extensions. Additionally, there are cases the housing adaptations would happen before the beginning of inhabitation, evidencing the process's complexity. The data collected show the different homeowners' pace for completing and improving their houses following diverse motivations in each stage. In this research, ten participant households (see Figure 1) were considered for the analysis and included the categorisation of residents into five groups:

- Steady performers
- Started then stalled
- Late starters
- Late occupants
- Low-density model performers

These classification criteria are based on the construction timeframes of extensions and examine the motivations and triggers and how the incremental construction became a continuous or intermittent process or even ignored for some time. The housing growth process is analysed based on the variations from late September 2013 when the homeowners received the houses, to July August 2017, when the data for this paper was collected. This analysis considers three components for the analysis of resident-controlled housing adaptations or extensions: (1) household size, (2) household income, and (3) physical changes in the initial house.

4.1. Steady Performers

This group of residents is analysed to know whether households that progressively performed changes in their houses represent the ideal of sustainability. In the two cases in Figure 2, residents regularly built extensions to the original house and completed more complex house additions.

House 01 (see Figures 1 and 2) household is a nuclear family of four members, and the father is working for Arauco. The family moved in immediately after they received their house and started to modify their home within the first year. The family decided first to build bedrooms in the upper floor extensions to accommodate the family comfortably. Later they built a grocery store on the ground floor, changing the use of living space, prioritising the family's need for extra income. This triggered the change in the space uses moving the kitchen to the back of the house and the dining and living room in front with independent access. The family income remained regular until 2016, when there was an increment, and they decided to cover the backyard because of the wind and the dust that mess up the laundry. The family expressed their satisfaction and pride about their house improvements because they financed with their savings. The residents mentioned their interest in keep modifying the house, specifically enlarging the living room by moving the toilet to the backyard. They also expressed their desire to expand the house vertically if this will be allowed in the future.

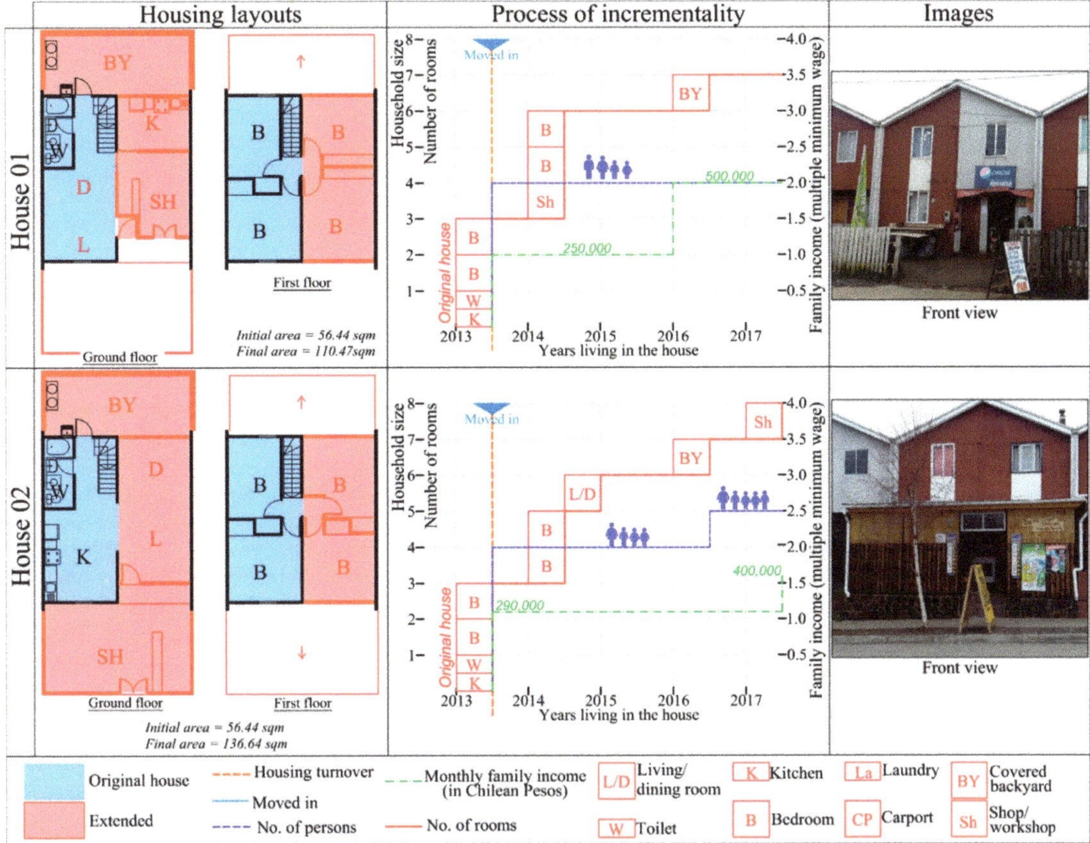

Figure 2. Housing extensions steadily built by residents. Source: Authors (2020).

Figure 2 also presents the process of change performed in house 02 (See Figure 1 to locate the house in the settlement) whose occupants are a single-parent family affected by the disaster. The mother is the head of the family with a daughter and a son, both adults and a grandson. The son and the daughter had various jobs in the city, and now the daughter runs a grocery shop in their home. The family moved into the new house immediately after they received it. Six months after the family moved in, they built bedrooms on the upper floor. Subsequently, the family built the living room and the dining room on the ground floor as specified by the house designers. All the extensions were built with the family savings as there was no increase in the monthly family income of 290,000 Chilean Pesos (approximately 482 US dollars), which is slightly higher than the monthly minimum wage in Chile (as of 2017). After three years of living in Villa Verde, the family decided to cover the backyard because it flooded during heavy rain. In the last year, the daughter's partner moved into the house and encouraged the family to open a grocery store financing its construction. Thus, the family income has increased, and the daughter can spend more time with her son. The family hired a carpenter to build the extensions; they feel their needs are covered and do not have plans to build more.

According to the cases shown above, residents who regularly build extensions were motivated by diverse factors such as the urgent need to accommodate the family members or the pressure to have an income source. The main reason to cover the backyard is to protect the family and the house from rain, wind and dust. Although the families' incomes are relatively low, they are willing to invest in improving their living conditions reflected in the housing extensions.

4.2. Started then Stalled

This group of residents promptly started to extend their houses, although they paused the housing growth process at one point. Understanding the residents' behaviour would explain the non-evident reasons behind the interruption or slowdown of building housing adaptations or extensions.

House 03 (Figure 3) belongs to a worker from a subcontractor of Arauco who moved in alone six months after he received the house. He only expanded in the house's upper floor with a carpenter's help, although it remains incomplete. In 2014, the owner received an unexpected income resulting from a former job's unemployment compensation that helped him finance the few extensions he built. Later he made few improvements, like installing ceramic tiles in the kitchen' flooring on the ground floor. The house remains unfinished because of the owner's lack of motivation until recently when the number of family members increased and became a nuclear family with two children. The owner is considering building extensions on the ground floor and finishings in the bedroom built upstairs.

The family living in house 04 (Figure 3) has initially four members, both parents and two children. The family was affected by the disaster and immediately moved into the new house after receiving it. In the following months, they expanded the house followed Elemental's design. However, the family changed the internal spaces' distribution motivated by privacy, which is the central family's priority. The family built a living room where initially the kitchen was located and moved the kitchen to the rear of the living room part of the extended half of the house on the ground floor. The two bedrooms built upstairs followed the designers' plans. There have not been considerable differences in the family size, the grandmother temporarily moved in to live with the family for six months in 2015. There was a slight increment in the household income as the father started working as a builder in a neighbouring housing project. The family selected the house at the end of the settlement because it is well concerned about privacy (Figure 1). However, since a new development is currently under construction, the family decided to build a front fence for safety reasons and define their property's limit. All the construction works have been done by themselves through self-construction and using the family savings. Although the family halted the house's construction, they expressed their intentions to pave the front garden and the backyard.

Residents interrupted the home extension construction when they considered that the house satisfied their needs. However, families might feel motivated to build new extensions or improvements when new needs emerge, whereas these might not be performed immediately. The second case presents motivations for change driven by the family's need for privacy and comfortable living spaces. Therefore, once residents satisfy their urgent needs, they might stop housing growth and focus on improvements of already-built spaces inside or outside their houses.

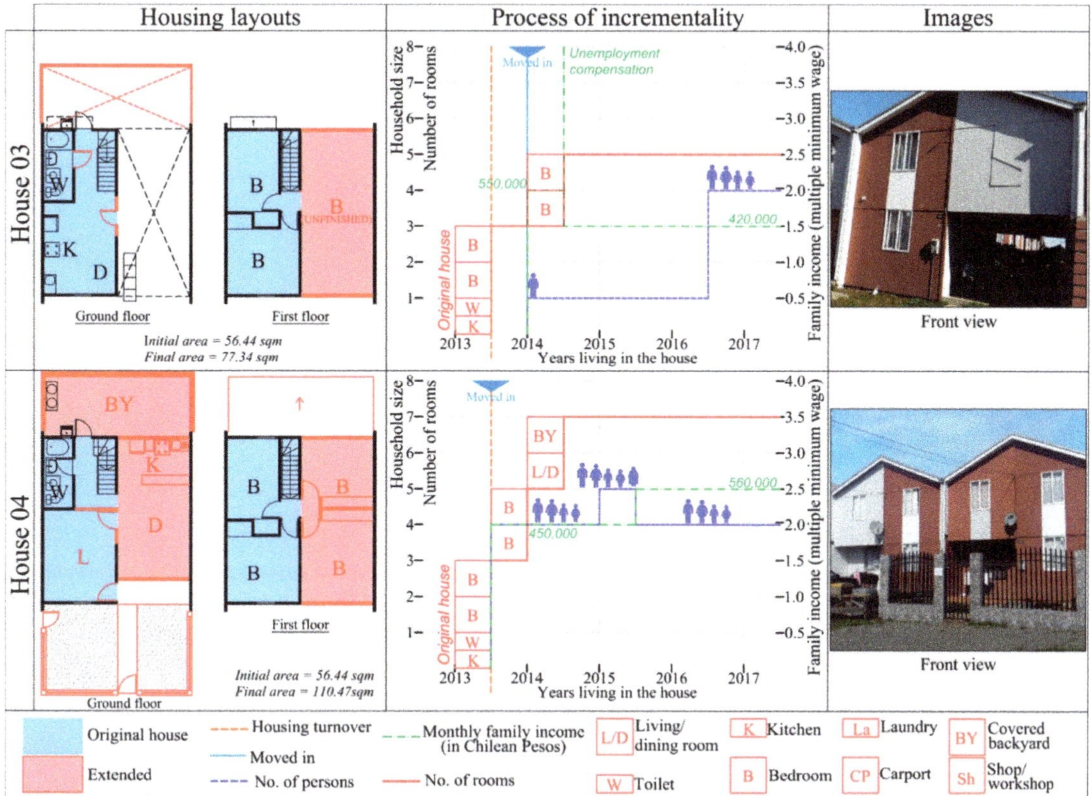

Figure 3. Residents that started and postponed the housing improvement. Source: Authors (2020).

4.3. Late Starters

Another group of residents delayed the construction of extensions, which despite the differences in the family size and type, at one point, they suddenly decided to start building. In the examples shown in Figure 4, residents experienced a long time of passiveness and decided not to extend their houses.

The family residing in house 05 (Figure 4) is an extended family, and its members increased since the homeowner, a retired lady, moved in alone in 2013. In early 2014, the homeowner's sister moved into the house, while their income remained slightly lower than the minimum wage. Almost three years later, the homeowners' daughter moved into the house with a consequent slight increase in the family income due to her occasional work. Finally, almost after four years of occupancy, the family decided to extend the house. The family decided to build an extra room with easy access for the daughter and her son when he was born. The family expressed their discontent for having most of their house unmodified and expect to build extensions on the ground floor in the short term. Their plans include building a dining and living room on the ground floor and one bedroom upstairs. Despite their incomplete house, the owner expressed her satisfaction with the house, especially the settlement location, accessible from the city centre about 20 to 30 min by foot. Although she also feels limited by the few options on what and how to build specified by Elemental.

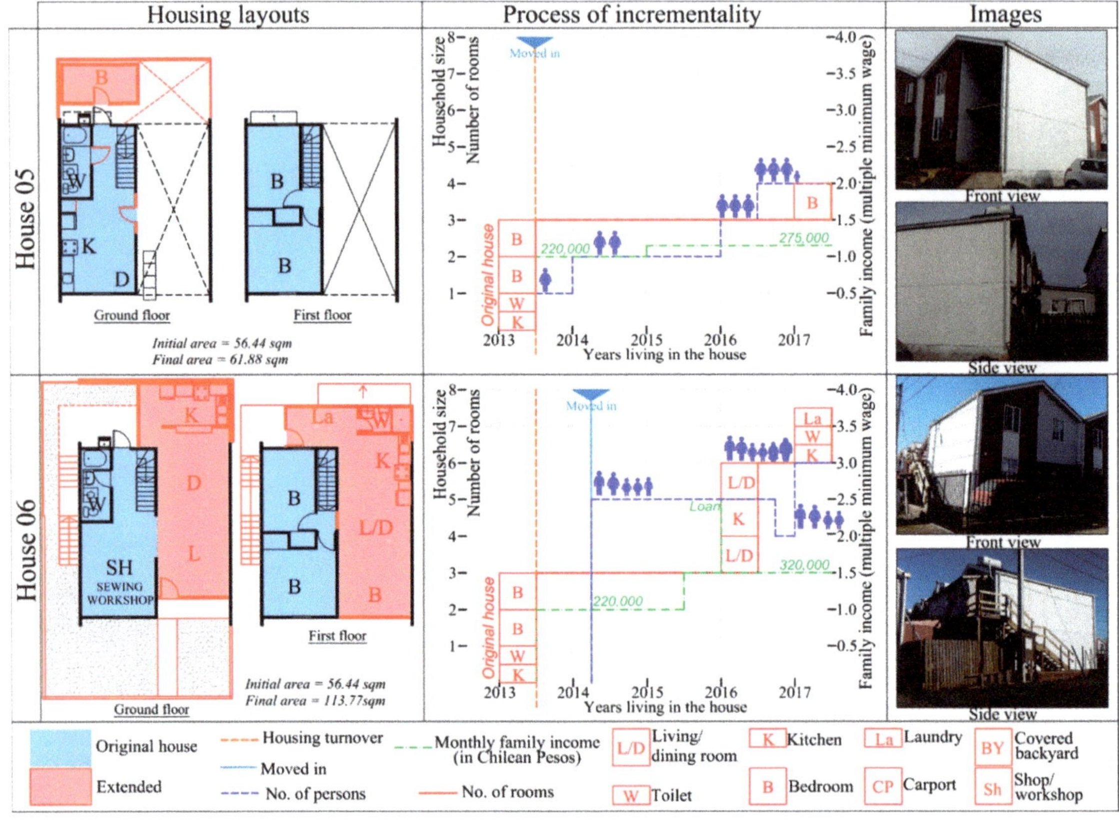

Figure 4. Delayed housing improvements. Source: Authors (2020).

The family living in house 06 (Figure 4) has five members with both parents and three children. The family former home was severely affected by the disaster and was awarded a house in Villa Verde. The family father is a full-time employee, and the mother is a dressmaker. The family delayed their occupancy, which moved in the first quarter of 2014, approximately six months after the housing turnover. The house remained unmodified until mid-2016, more than two years after the family moved in. The reasons to extend the home were related to the need to satisfy the family's space needs and provide them with a comfortable living environment that was finally possible thanks to a bank loan applied by the family. The family also changed the uses of the spaces. The backyard was covered and the kitchen was moved there. In this way, the house's frontal part could be converted into a sewing workshop for the family's mother to establish a home business. All these works were done through self-construction with relatives' support. Later, a second event triggered unexpected renovations and extension of the house. In January 2017, there were wildfires in the forest areas in the region, and the parents' house of the family father, located in a rural area near the city of Constitución, resulted heavily damaged. Suddenly, the family needed to modify the house again. They built a separate entrance to the upper floor, a space for laundry and toilet, and renovated two bedrooms' space to make a multipurpose space for a studio apartment. In the future, the family wants to make internal renovations and applied for government subsidies through the Home and Neighbourhood Improvement Program, but the outcome is delayed.

In the previous examples, at first, residents remained passively living in the original house. In both cases, the residents' expressed their unconformity for not extending their house earlier and compared their homes with their neighbours' wishing to actively improve their living conditions and have comfortable spaces for their families. Although there were financial constraints, the sudden change in the family size motivated a rapid change and expansion of the houses. Both homeowners expressed their will to continue improving their homes in the future.

4.4. Late Occupants

The fourth group identified is the residents that delayed their moving into the houses. In fact, in the survey conducted in Villa Verde, 84 percent of the households interviewed delayed occupying their houses. The 68 percent did it because they wanted to build the extensions before moving in, and 16 percent lacked the motivation to extend. Thus, most of the residents that occupied the houses late felt that they needed to build first

The family residing in house 07 (Figure 5) moved to Villa Verde one year after receiving their house. The parents expressed that it was difficult to live in an incomplete house with the whole family. The housing extensions included the construction of the living room and two bedrooms on the upper floor. Additionally, the change of use of one of the original bedrooms and converted it into a toilet. The priority was to have a comfortable living environment for the four family members when they moved into the house. Later, the family covered the backyard to protect the house from heavy rain and wind. In the next year, the daughter moved out and later the son. Although in 2015, there was a particular income from the father's unemployment insurance, he did not invest in renovating or improving their home. Recently, residents built a low fence to define their lot and paved the house entrance and worked on their frontal garden. Nowadays, the parents are a retired couple who does not have plans for building further housing extensions.

The family residing in house 08 (Figure 5) was granted a house in Villa Verde because they lost their home due to the disaster. The family moved in Villa Verde house in 2015, about one and a half years after receiving their house, because they wanted to build the extensions first. The household is a nuclear family type with both parents and three sons. The initial extensions included completing the other half of the house, such as the living room on the lower floor and bedrooms upstairs. The kitchen was also renovated to make space for a small eatery. The family covered the carport to protect their cars because the father works in transport. The family feels proud and satisfied with the house as it results from their efforts and is highly motivated to continue with housing renovations and extensions.

The cases presented show the residents' concerns about building housing extensions to provide comfortable living environments for the family before moving in. The residents are proud of the works performed in their homes and feel attached to them. Families are motivated to renovate some existing spaces and build more extensions in the future.

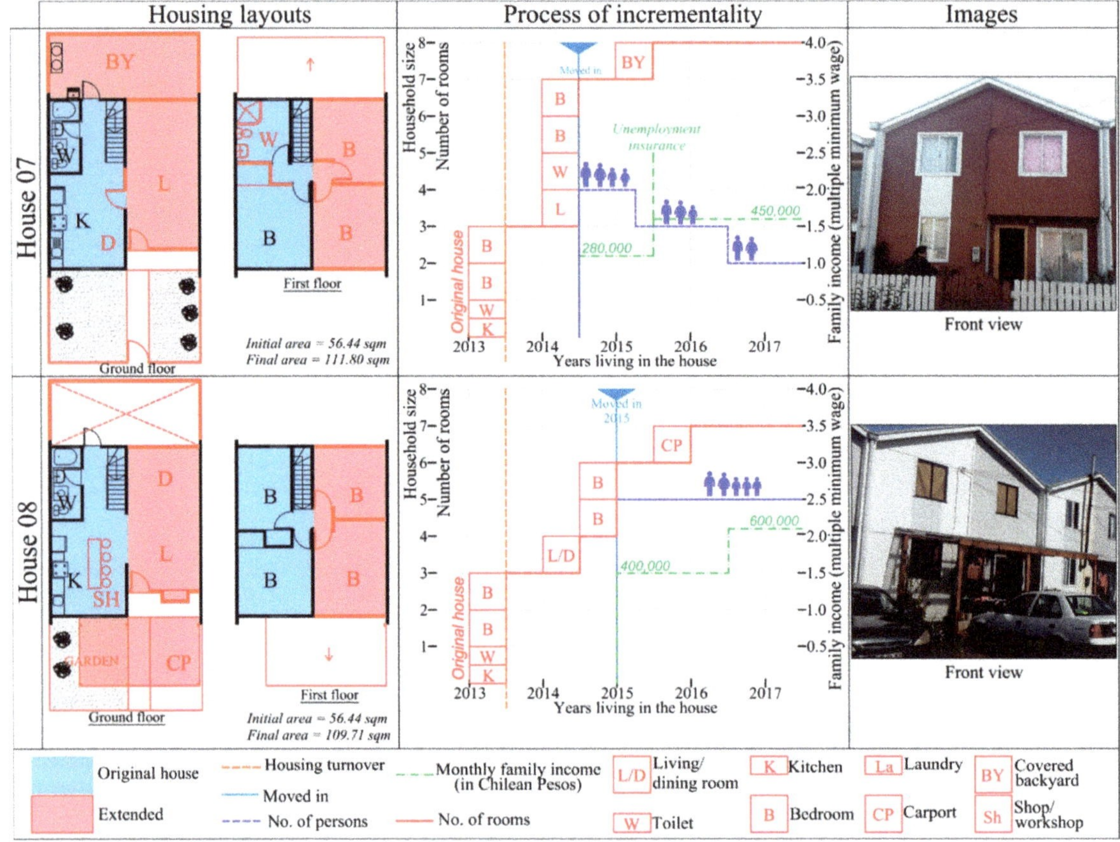

Figure 5. Late occupants and progressive housing construction. Source: Authors (2020).

4.5. Low-Density Model Performers

The last group analysed are the small-sized households that performed multiple extensions and improvements, although the initial house would have been enough to enjoy comfortable habitable spaces. Both households were affected by the disaster. Due to the household size, building housing extensions might are not directly related to the urgency to provide a quality of life to the residents.

The owner of house 09 (Figure 6) is a senior woman who moved in with her adult daughter, who moved in during the first year. The homeowner receives a modest income from her retirement pension and additional casual income from baking pastries. They delayed the construction of housing extensions for almost two years. The homeowner's daughter financed most of the housing improvements to provide a comfortable living environment for her mother. A carpenter who is a neighbour in Villa Verde built the extensions and renovations. Most of the construction works were done simultaneously. The internal renovations include reducing the living room to build a bedroom for the mother anticipating her future reduced mobility due to her advanced age. The bedrooms on the upper floor were built to accommodate the homeowner's daughter and son during their visits. The backyard was partly covered and used as storage. The owner feels that the house improvements satisfy her needs. However, she is coordinating with the Villa Verde senior's association to apply for government subsidies to renovate the kitchen and toilet.

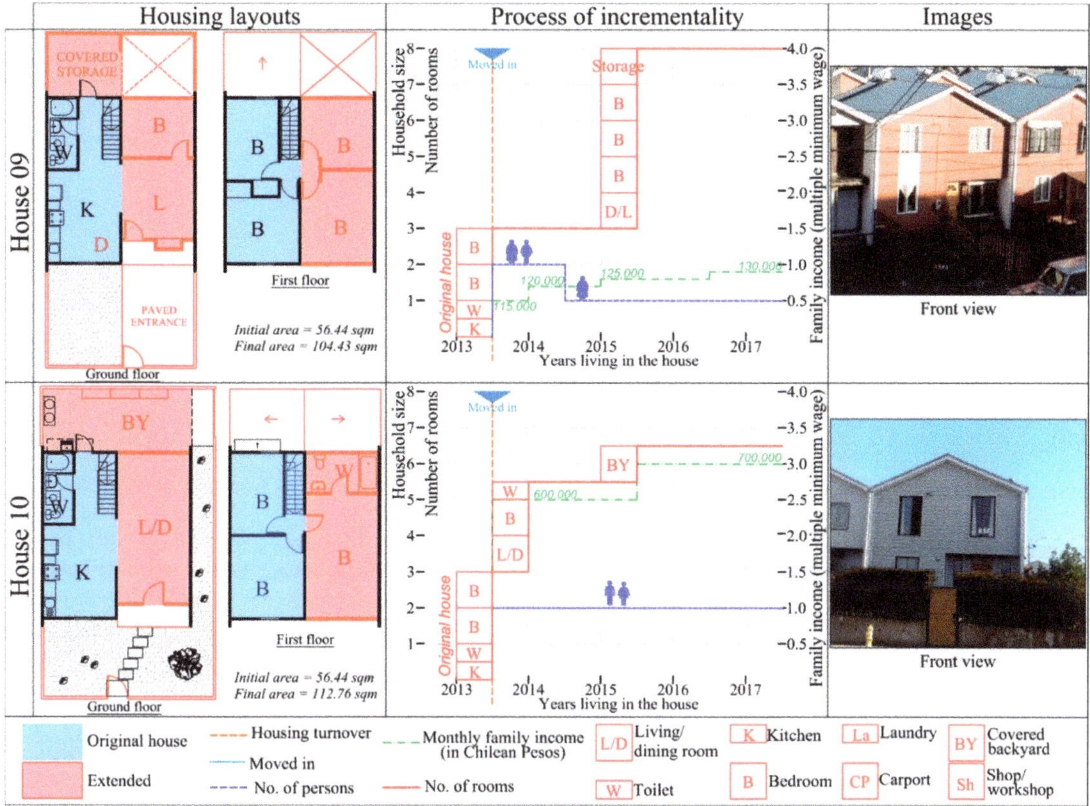

Figure 6. Small size households as model incremental builders. Source: Authors (2020).

The family residing in house 10 (Figure 6) is a childless couple with a higher family income as both work full time, the wife as a hairdresser and the husband as a taxi driver. They were affected by the 2010 earthquake and tsunami and received the house as disaster victims. The wife got actively involved in the planning of the houses and later was a community leader. Thus, she got the priority for selecting her house's location in Villa Verde. The initial improvements in the house were the planned extensions built just after they moved in. They used their family savings before they received the house. One of the initial improvements included constructing a toilet for the master room on the upper floor and the other two bedrooms have flexible uses. The family also expanded the limits of their lot to the house's side, which they coordinated with the neighbours. Later they covered the backyard to use it as laundry space, storage, and a small-scale hydroponics urban farm as the residents practice a "healthier lifestyle". The family built a fence and worked in the frontal and side garden. Their plans for future housing improvement include protective structures against rain like a covered entrance.

In previous cases, residents were initially motivated to satisfy their space needs. However, later other motivations emerged based on the residents' lifestyles and special needs.

5. Discussion: A Resident-Controlled Process

The analysis of the progress in the construction of housing extensions between disaster-affected and non-affected households (see Figure 7 and Table 1) suggest that both groups were in similar conditions at the time they moved into their houses in Villa Verde. Residents used the three-year gap between the disaster and the housing turnover to prepare emotionally and financially to build extensions. Thus, the information suggests that residents have similar capacities to modify when the circumstances motivate them to build. For instance, when they have sufficient financial resources or when they feel pressured by family needs.

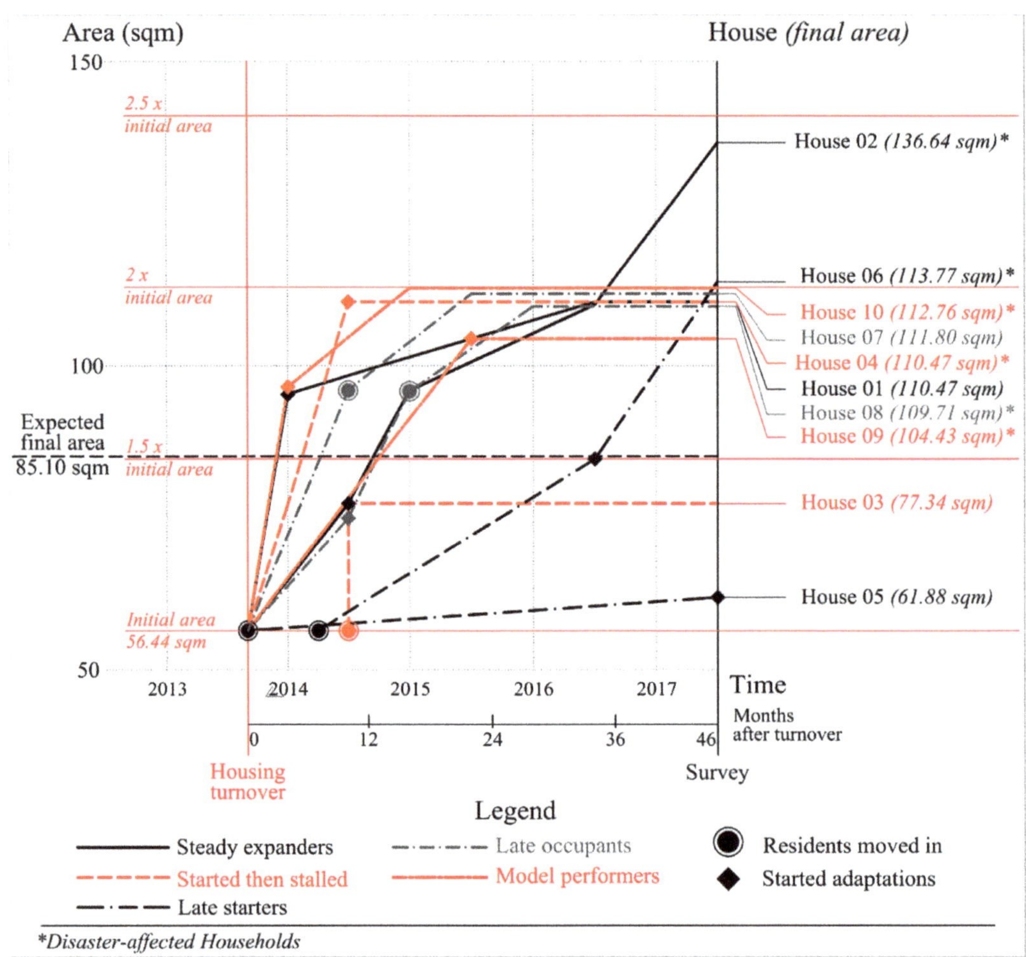

Figure 7. Temporality and progress of housing modifications in Villa Verde. Source: Authors (2020).

Looking at the timeframes and the progress of housing extensions as indicator of self-recovery, this paper observes a variable increase in the floor area from minimum to more than double the initial house designed by Elemental, as seen in Figure 7. Remarkably, the designers' expected final area of 85.10 sqm was rapidly achieved by 7 out of 10 households within the first year after receiving their houses with disaster-affected households among the better performing residents. Furthermore, the major construction works were performed within the first 15 months. Moreover, 6 out of the 10 households studied added between 50 to 100 percent to the initial house area, and two households more than doubled the initial house area. Figure 7 also shows the various paces of the housing modifications that have been identified in the five groups of residents. However, there is no clear evidence to link the housing modification's pace and the quantity of area added to the houses. A closer look at the periods of active construction presented in Table 1 shows no clear pattern in activity level among the five groups of households within the four-year timeframe despite the different housing construction paces. However, there were slightly less active periods in the group of residents that paused the constructions and those who moved in later. Conversely, looking at the area built during active periods (semesters), we observe that the less active households and the most active added similar areas to the original house.

Table 1. Progress of housing adaptations and areas.

Group	House No.	Periods of Active Construction *	Final Area **	Area Added **	Area Built during Active Semesters **	Beyond Designers' Framework	Future Extensions/Renovations	Built Internal Renovations	Applied for Subsidies †
Steady performers	House 1	2	110.47	54.03	27.02	x	x	x	
	House 2※	4	136.64	80.20	20.05	x			
Started then stalled	House 3	1	77.34	20.90	20.90		x		
	House 4※	2	110.47	54.03	27.02	x	x	x	
Late starters	House 5	1	61.88	5.44	5.44		x		
	House 6※	2	113.77	57.33	28.67	x	x	x	x
Late occupants	House 7	2	111.8	55.36	27.68	x			
	House 8※	3	109.71	53.27	17.76	x	x	x	
Low-density model performers	House 9※	2	104.43	47.99	24.00	x	x	x	x
	House 10※	4	112.76	56.32	14.08	x	x	x	

Notes: ※ Disaster-affected households. * Measured in semesters based on interviewees' preferences to define timeframes. ** In square meters. † Subsidies must be collective and only in areas previously approved in the building permit.

This study also shows that the construction pace has no impact on where the residents-built extensions as residents in all the groups built beyond the designers' framework. Furthermore, all except two expressed their motivation to keep building in the future, and most of them already have plans of what they want to build. Even though most of the discussion in this paper was related to the housing extensions and self-recovery, residents also invested in improving the spaces already built. We observe that six households built internal renovations, including changing the houses' original layout or replacing specific elements like renovating the kitchens and toilets or adding finishing to the houses.

This study also observes that few households applied for government-funded subsidies for internal renovations due to the Housing and Neighbourhood program conditions, limiting the eligibility to collective applicants. In Villa Verde, one successful collective application was submitted by the seniors' neighbours association and the second was in preparation by one of the neighbours' associations and targeted the poorer households and the large families. However, they often complained about the limited control over the government-funded renovations that contractors performed.

Residents built extensions at different paces and following various motivations. Despite the differences, this study observed that Villa Verde houses' incremental construction is sustainable in time and is affected by the family conditions changes. For instance, residents prepared for building their houses before moving in, and investing in construction was their priority. Homeowners' urgency to accommodate family members motivated the construction of extensions. Residents modified their houses due to changes in household conditions and size, which sometimes triggered unexpected extensions and renovations. However, the houses' changes were also motivated to change their financial or family conditions. For instance, the construction of grocery shops and home-based businesses affected the families' financial situation, which might motivate residents to build more extensions or renovate the already-built spaces. As we observed during the interviews with residents, building and improving their houses is part of their daily lives as they expressed their pride in the extensions performed and are highly motivated to keep building.

This study has presented the outcomes of facilitating housing frameworks for resident-controlled incremental construction and self-recovery to empower disaster-affected people to take care of their own recovery. Twigg [44] highlights building back safer as paramount in the self-recovery discourse that focuses on avoiding the recreation of unsafe designs and building practices. For instance, the seismic performance of the Quinta Monroy project, also designed by Elemental, was tested during the 2014 Iquique Earthquake. The positive outcomes resulted from the residents' willingness to use only lightweight materials, which the designers encouraged in their technical manuals for building extensions [22]. However, the impacts of the complexity in the types of extensions built beyond the designers' technical specifications in Villa Verde remain unclear. Therefore, it is necessary to evaluate the long-term implications of the Elemental's incremental housing model in the residents' safety and quality of their living conditions which will define opportunities for improvement and replicability in other contexts.

6. Conclusions

This study presented different patterns for housing adaptations analysed almost four years after Villa Verde's houses were given to the families. Capturing the evolution of incremental housing construction evidenced complexities resulting from the variety of households' characteristics and needs in their post-disaster housing self-recovery.

Unlikely strictly following Elemental's plans for extensions, residents presented different alternatives and reasons beyond the designer's manual regardless of their disaster-affected or unaffected household condition. Residents started building at different times and at different paces based on regularly emerging needs or sudden family income and size changes. The complexity of unpredictable, driven factors for housing adaptations is shown in some families' need to create an income source. In other cases, it is to accommodate the diversities of family compositions or simply adapt their home to suit their emerging spatial requirements. Therefore, the unpredictability of the changing households' conditions and behaviours defines the unexpected changes in the houses that often-covered areas beyond the designer's parameters. Houses in Villa Verde were designed to be changed, although the design allowed only specific future extensions trying to anticipate unpredictable needs. This paper evidenced that despite "pre-packaged" alternatives for housing adaptations, providing a "framework" that supports the housing extensions may challenge the conventional designers' ideals leading to what Leupen (2006) calls "designing for the unknown".

Aravena and Elemental tried to provide a platform for resident-controlled housing development, and the Chilean regulations also acknowledge and support incremental housing. Remarkably, vulnerable residents accessed government funding for housing renovations, although these were limited to internal spaces and controlled by contractors. However, for most of the housing extensions, designers and the government failed to try to limit a continuous and unpredictable process whose implications on appropriate living conditions and structural stability remain uncertain. In this context, designers and policymakers need to understand that housing is a mean to provide residents the freedom to be and exercise their rights and take responsibility for the needed decisions to fulfil specific aspirations. Therefore, once residents were given the freedom to modify their houses for an unpredictable time and in unexpected ways that show a "freedom of the unknown". This study evidenced residents' capacities and dedication to self-recover by controlling the incremental housing construction process. Accordingly, long-term analysis of post-disaster incremental housing is crucial to minimise the creation of further risks, which requires community engagement in creating safer living environments.

Author Contributions: Conceptualization, S.C.; Investigation, D.O.; Methodology, S.C.; Supervision, D.O.; Visualization, S.C.; Writing—Original draft, S.C.; Writing—Review and editing, D.O. All authors have read and agreed to the published version of the manuscript.

Funding: This research received no external funding.

Institutional Review Board Statement: Research ethics approval for this project has been awarded by The University of Melbourne Ethics Committee for the project "Resident-initiated housing adaptations in projects designed under Incremental Housing principles in Chile", Ethics ID number 1749143.1.

Informed Consent Statement: Verbal informed consent was obtained from all subjects involved in the study.

Data Availability Statement: The quantitative dataset for the project "Resident-initiated housing adaptations in projects designed under Incremental Housing principles in Chile" is available at https:

//doi.org/10.48331/scielodata.J6KE9P/CBUMEN, SciELO Data, V1, UNF:6:9XTsklsgaBJfJJ3LLaK38g==
[fileUNF] (accessed on 25 July 2022). Unprocessed interview data such as participants' transcripts are susceptible to the project's ethics concerns and cannot be disclosed as protecting the identifiable information is a higher concern in respecting participants' privacy and confidentiality under the Australian National Statement on Ethical Conduct in Human Research (2007) updated in 2018 https://www.nhmrc.gov.au/about-us/publications/national-statement-ethical-conduct-human-research-2007-updated-2018 (accessed on 25 July 2022).

Acknowledgments: The authors would like to thank the residents of Villa Verde for their generous participation in this research program and Julio Carrasco (B. Arch.) for his valuable support during the collection of information.

Conflicts of Interest: The authors declare no conflict of interest.

References

1. McGlade, J.; Bankoff, G.; Abrahams, J.; Cooper-Knock, S.; Cotecchia, F.; Desanker, P.; Erian, W.; Gencer, E.; Gibson, L.; Girgin, S. *Global Assessment Report on Disaster Risk Reduction 2019*; UN Office for Disaster Risk Reduction: Geneva, Switzerland, 2019.
2. Sharma, A. Supporting Locally Driven Shelter Responses. In *The State of Humanitarian Shelter and Settlements 2018*; Global Shelter Cluster, Ed.; Global Shelter Cluster: Geneva, Switzerland, 2018; pp. 19–24.
3. ShelterBox. *Climate Crisis to Destroy 167 Million Homes in Next 20 Years*; ShelterBox: Truro, UK, 2021.
4. Carrasco, S.; O'Brien, D. Urbanism of Emergency: Use and Adaptation of Public Open Spaces in Disaster-Induced Resettlement Sites. In *Resettlement Challenges for Displaced Populations and Refugees*; Springer: Berlin/Heidelberg, Germany, 2019; pp. 163–174.
5. Oliver-Smith, A.; de Sherbinin, A. Resettlement in the twenty-first century. *Forced Migr. Rev.* **2014**, *45*, 23–25.
6. Peacock, W.G.; Dash, N.; Zhang, Y.; Zandt, S.V. Post-Disaster Sheltering, Temporary Housing and Permanent Housing Recovery. In *Handbook of Disaster Research*; Springer: Cham, Switzerland, 2018; pp. 569–594.
7. Kroll-Smith, S.; Baxter, V.; Jenkins, P. *Left to Chance: Hurricane Katrina and the Story of Two New Orleans Neighborhoods*; University of Texas Press: Austin, TX, USA, 2015.
8. Ahrentzen, S.B. Home as a Workplace in the Lives of Women. In *Place Attachment*; Altman, I., Low, S.M., Eds.; Springer: Berlin/Heidelberg, Germany, 1992; pp. 113–138. [CrossRef]
9. Carrasco, S.; Ochiai, C.; Okazaki, K. Residential satisfaction and housing modifications: A study in disaster-induced resettlement sites in Cagayan de Oro, Philippines. *Int. J. Disaster Resil. Built Environ.* **2017**, *8*, 175–189. [CrossRef]
10. Flinn, B.; Schofield, H.; Morel, L.M. The case for self-recovery. *Forced Migr. Rev.* **2017**, *55*, 12–14.
11. Jha, A.K.; Barenstein, J.D.; Phelps, P.; Pittet, D.; Sena, S. *Safer Homes, Stronger Communities: A Handbook for Reconstructing after Natural Disasters*; World Bank Publications: Herndon, VA, USA, 2010.
12. Schofield, H.; Lovell, E.; Flinn, B.; Twigg, J. Barriers to urban shelter self-recovery in Philippines and Nepal: Lessons for humanitarian policy and practice. *J. Br. Acad.* **2019**, *7*, 83–107.
13. Twigg, J.; Lovell, E.; Schofield, H.; Miranda Morel, L.; Flinn, B.; Sargeant, S.; Finlayson, A.; Dijkstra, T.; Stephenson, V.; Albuerne, A. *Self-Recovery from Disasters: An Interdisciplinary Perspective*; Overseas Development Insitute: London, UK, 2017.
14. Ahmed, I.; Parrack, C. Shelter Self-Recovery: The Experience of Vanuatu. *Architecture* **2022**, *2*, 434–445. [CrossRef]
15. Ferguson, B.; Smets, P. Finance for incremental housing; current status and prospects for expansion. *Habitat Int.* **2010**, *34*, 288–298. [CrossRef]
16. Nohn, M.; Goethert, R. *Growing Up! The Search for High-Density Multi-Story Incremental Housing*; SIGUS-MIT & TU Darmstadt: Darmstant, Germany, 2017.
17. UN-Habitat. Affordable land and Housing in Latin America and the Caribbean. In *Adequate Housing Series*; United Nations Human Settlements Programme, UN-Habitat: Nairobi, Kenya, 2011; Volume 1.
18. Turner, J. The squatter settlement: An architecture that works. *Archit. Des.* **1968**, *38*, 355–360.
19. Arriagada, C.; Sepúlveda Swatson, D.; Cartier Rovirosa, E.; Gutiérrez Vera, C. *Chile: Un Siglo de Políticas en Vivienda Y Barrio [Chile; a Century of Housing and Neighbourhood Policies]*; Ministerio de Vivienda y Urbanismo-MINVU: Santiago, Chile, 2004.
20. Turner, J.F.; Fichter, R. *Freedom to Build: Dweller Control of the Housing Process*; Macmillan: New York, NY, USA, 1972.
21. Habraken, N.J. *Supports: An Alternative to Mass Housing*; Architectural Press: London, UK, 1972.
22. Carrasco, S.; O'Brien, D. Beyond the freedom to build: Long-term outcomes of Elemental's incremental housing in Quinta Monroy. *Urbe. Rev. Bras. Gest. Urbana* **2021**, *13*. [CrossRef]
23. Negro, V. Arquitectura Participativa en America Latina [Participatory architecture in Latin America]. *Casa Del Tiempo* **2016**, *3*, 42–45.
24. Vergara Perucich, F.; Boano, C. Bajo escasez ¿Media casa basta? Reflexiones sobre el Pritzker de Alejandro Aravena {In scarcity Is half-the-house enough? Reflections on the Alejandro Aravena's Pritzker}. *Rev. Arquit.* **2016**, *21*, 37–46. [CrossRef]
25. Arauco. *Celulosa Arauco Y Constitución S.A. Memoria Anual 2010*; ARAUCO: Santiago, Chile, 2011.
26. O'Brien, D.; Carrasco, S. Chapter 12-Incremental Housing in Villa Verde, Chile: A View through the Sendai Framework Lens. In *Enhancing Disaster Preparedness*; Martins, A.N., Fayazi, M., Kikano, F., Hobeica, L., Eds.; Elsevier: Amsterdam, The Netherlands, 2021; pp. 223–240. [CrossRef]

27. Davis, I.; Parrack, C. Taking the Long View. In *The State of Humanitarian Shelter and Settlements 2018*; Global Shelter Cluster, Ed.; Global Shelter Cluster: Geneva, Switzerland, 2018; pp. 9–14.
28. Massachussets Institute of Technology (MIT). Three Levels of Surveys are Anticipated in a Full Understanding of Incremental Housing. 2010. Available online: http://web.mit.edu/incrementalhousing/understandingFormat/index.html (accessed on 22 January 2022).
29. Gattoni, G.; Goethert, R.; Chavez, R. *Self-Help and Incremental Housing El Salvador: Likely Directions for Future Policy.*; Massachusetts Institute of Technology, Fundación Salvadoreña de Vivienda Minima (FUNDASAL), Universidad José Simeón Cañas (UCA), Universidad de El Salvador (UES): San Salvador, El Salvador, 2011.
30. Goethert, R. Capturing Process of Informal Housing Development: A Longitudinal Survey Methodology, A Pattern Recognition Approach. In Proceedings of the Un World Urban Forum 7, Medellin, Colombia, 5–11 April 2014.
31. Yin, R.K. *Case Study Research: Design and Methods*; Sage Publications: Los Angeles, CA, USA, 2013.
32. Zeisel, J. *Inquiry by Design: Tools for Environment-Behaviour Research*; Cambridge University Press: Cambridge, UK, 1984.
33. Groat, L.N.; Wang, D. *Architectural Research Methods*; John Wiley & Sons: Hoboken, NJ, USA, 2013.
34. Taylor, R.; Forrester, J.; Pedoth, L.; Matin, N. Methods for Integrative Research on Community Resilience to Multiple Hazards, with Examples from Italy and England. *Procedia Econ. Financ.* **2014**, *18*, 255–262. [CrossRef]
35. Faber, M.H.; Giuliani, L.; Revez, A.; Jayasena, S.; Sparf, J.; Mendez, J.M. Interdisciplinary Approach to Disaster Resilience Education and Research. *Procedia Econ. Financ.* **2014**, *18*, 601–609. [CrossRef]
36. Lombard, M.; Meth, P. Informalities. In *Urban Theory*; Routledge: London, UK, 2016; pp. 180–193.
37. Aravena, A.; Iacobelli, A. *Elemental: Manual de Vivienda Incremental Y Diseño Participativo [Elemental: Incremental Housing and Participatory Design Manual]*; Hatje Cantz: Ostfildern, Germany, 2016.
38. MINVU.; Municipio de Constitución; Arauco. *Plan de Reconstrucción Sustentable PRES Constitución [Plan for the Sustainable Reconstruction of Constitucion]*; Ministerio de Vivienda Y Urbanismo: Santiago, Chile, 2010.
39. MINVU. *Texto Actualizado Del Decreto Supremo N° 174, (V. Y U.), de 2005 d.o. de 09.02.06 Reglamenta Programa Fondo Solidario de Vivienda*; Habitacional, D.P., Ed.; Ministerio de Vivienda y Urbanismo [Ministry of Housing and Urban Planning]: Santiago, Chile, 2005.
40. MINVU. *LEY 20741-Modifica la Ley N° 19.537, Sobre Copropiedad Inmobiliaria, Para Facilitar la Administración de Copropiedades Y la Presentación de Proyectos de Mejoramiento O Ampliación de Condominios de Viviendas Sociales*; Ministerio de Vivienda y Urbanismo [Ministry of Housing and Urban Planning]: Santiago, Chile, 2014.
41. MINVU. *D.S No 27 Reglamento del Programa de Mejoramiento de Viviendas Y Barrios*; Ministerio de Vivienda y Urbanismo [Ministry of Housing and Urban Planning]: Santiago, Chile, 2018.
42. Friedman, A. *The Adaptable House*; McGraw-Hill, Inc.: New York, NY, USA, 2002.
43. Leupen, B.; Heijne, R.; van Zwol, J. *Time-Based Architecture*; 010 Publishers: Rotterdam, The Netherlands, 2005.
44. Twigg, J. The Evolution of Shelter "Self-recovery": Adapting Thinking and Practice for Post-disaster Resilience. *J. Br. Acad.* **2021**, *9*, 5–22. [CrossRef]

Article

Design and Disaster Resilience: Toward a Role for Design in Disaster Mitigation and Recovery

Esther Charlesworth * and John Fien 

School of Architecture and Urban Design, RMIT University, GPO Box 247V, Melbourne 3001, Australia; john.fien@rmit.edu.au
* Correspondence: esther.charlesworth@rmit.edu.au; Tel.: +61-410560411

Abstract: This paper examines how the discourses and practices of design can be applied to both mitigate the damaging impacts of (un-)natural disasters and guide resilient post-disaster recovery. Integrated with systems analysis, design can provide both an innovative window for understanding the complexities of disaster-risk reduction and recovery, as well as a conceptual bridge to new ways of building socio-economic and physical resilience in disaster-affected communities. However, the skills of key systems and design thinkers, such as architects, urban planners, and landscape architects, are seldom employed, despite their demonstrated capacity to work with disaster-prone or -impacted communities to develop integrated spatial responses to guide both disaster-risk reduction and long-term rebuilding after a disaster. Indeed, there has been little focused investigation of the potential contributions of design per se in developing strategies for disaster-risk reduction and recovery. Similarly, there has been little attention in design education to complementing the creative problem-solving skills of the designer with the contextual and systemic understandings of disaster management and disaster-resilient design. This paper addresses these omissions in both disaster management and design education though a review of research on design contributions to disaster issues and provides a case study of the curriculum and pedagogical approaches appropriate to build capacity for enhancing this contribution.

Keywords: disaster mitigation; disaster-risk reduction; disaster recovery; design; architecture; design thinking

Citation: Charlesworth, E.; Fien, J. Design and Disaster Resilience: Toward a Role for Design in Disaster Mitigation and Recovery. *Architecture* **2022**, *2*, 292–306. https://doi.org/10.3390/architecture2020017

Academic Editor: Iftekhar Ahmed

Received: 25 February 2022
Accepted: 31 March 2022
Published: 13 April 2022

Publisher's Note: MDPI stays neutral with regard to jurisdictional claims in published maps and institutional affiliations.

1. Introduction

This paper examines how the discourses and practices of design can be applied to both mitigate the damaging impacts of (un-)natural disasters before they occur and guide effective post-disaster recovery. From annual cyclones in the tropics and sub-tropics, to earthquakes, floods, heat stress, and wildfires, the world is experiencing an increasing number of disasters, which have caused damage in excess of USD 5200 billion since 1980, and USD 150 billion in 2019 alone (Munich Re 2020) [1]. The intensity and cascading nature of disasters have doubled the number of people displaced in the past two decades: a rate of greater than one person per second, every day, every year. The 1900 disasters reported in 2019 displaced 24.9 million people from their homes, globally. This figure is the highest recorded since 2012 and three times the number of displacements caused by conflict and violence (IDMC 2020a) [2]. Flooding, alone, is predicted to cause 50 million displacements a year by 2100 if climate change is not addressed (IDMC 2020b) [3].

However, the scale of these losses is not the only or, indeed, the major problem. Thus, Cadman (2020) [4] argues, *the key issue is determining what needs to be done to make communities and their homes and infrastructure more resilient.* However, government and community agencies often struggle to implement effective strategies to prevent or mitigate the impacts before such disasters occur and to plan for long-term recovery afterwards (Bojic, Baas, and Wolf 2019) [5]. The central challenge these agencies face is the development of policies and practices for reducing the vulnerabilities that prevent communities from becoming more disaster-resistant (Geis 2000) [6] or -'resilient' (Aldunce et al., 2014 [7], Alexander 2013 [8]).

Increasingly, scholars are identifying whole-of-system risks related to the design of the built environment—including unsafe settlement patterns and the inappropriate design of buildings in disaster-prone areas—as key causes of the vulnerabilities that need to be addressed (Fisher 2013) [9]. These studies include, for example, inquiries into wildfire damage (e.g., Gonzalez-Mathiesen and March 2018) [10] and the impacts of flooding and cyclones (e.g., Smith and Low Choy 2014) [11]. Similarly, a review of reconstruction projects after the 2004 Boxing Day tsunami in Sri Lanka found that poor urban design was responsible for many villages being rebuilt in unsafe locations and lacking in basic infrastructure for water, sewerage, and electricity (Ahmed and Charlesworth 2015) [12].

While many such housing and infrastructure challenges are critically related to design issues, design also provides both an innovative window for understanding the complexities of disaster-risk reduction and recovery, as well as a conceptual bridge to new ways of building socio-economic and physical resilience in disaster-affected communities (Keenan 2018) [13]. Predominant approaches to complex disaster events tend to focus on individual elements of regional or urban systems, such as constructing fire breaks or flood levees (Lee, 2020) [14]. As such, they are often not well-suited to mitigating the original system vulnerabilities that can turn extreme weather events into major disasters. This is because the complexity of such systemic problems requires approaches to resolution that recognize systemic connections (in which the solutions to today's problems may become tomorrow's new problem) and the consideration of creative and diverse alternatives. This is the realm of design thinking because it helps tackle complex and uncertain challenges which are commonly seen as 'wicked problems'. In the increasingly complex and uncertain world of disaster-risk management, design thinking avoids linear thinking and, instead, looks to two iterative processes: (i) identifying and formulating the problems by understanding systemic relationships, and then (ii) developing and testing alternative solutions. The first is an analytic sequence in which the designer investigates all the social, cultural, environmental, political, and economic elements and interconnections within related systems. This is a key task as a poorly formulated problem will not lead to a lasting solution. The second involves sequences of exploring, identifying, and testing several possible approaches to specify best-possible resolutions for the problem. (The design thinking process in relation to wicked problems has been elaborated in a wide range of publications since Buchanan (1992) [15]. See, for example, (Johnson 2016 [16])).

However, the skills of key systems and design thinkers (such as architects, urban planners and designers, and landscape architects) are seldom employed in disaster mitigation and recovery despite their demonstrated capacity to work with communities and to develop integrated spatial responses to guide both disaster-risk reduction and long-term rebuilding after a disaster (Charlesworth 2014) [17]. Indeed, despite the growing body of research on *urban planning* in relation to risk and climate-change adaptation (e.g., Elliot-Ortega 2010 [18], Fisher 2013 [9], Shaw, Rahman, Surjan, and Parvin 2016 [19] American Society of Landscape Architects 2020 [20]), there has been little focused investigation of the potential contributions of *design* in developing strategies for disaster-risk reduction and recovery with and for impacted communities. Similarly, there has been little attention in design education to complementing creative problem-solving skills of the designer with the contextual and systemic understandings of disaster management or disaster-resilient design. Thus, it is understandable that design expertise is not often applied in recovery and reconstruction as the number of architects equipped to respond in such situations is still very low.

This paper seeks to address this critical gap in both knowledge and practice by focusing on two questions:

(i) What can be learnt from examples of design-based responses to disaster mitigation and recovery?

(ii)   How can the skills needed to integrate design into disaster-risk management be developed?

A theoretical foundation for answering these questions is provided in the next section. Then, the paper turns to the first question in a section that explores several examples of design-based responses to disaster. This section is based upon a cross-case analysis of case studies of 'humanitarian architecture' conducted by Charlesworth (2014) [17]. A key finding from the cross-case analysis was the lack of attention to providing design strategies for disaster and conflict scenarios in the professional education of the architects involved. Thus, the final section of the paper investigates ways in which these skills could be included in future design education through a brief review of current postgraduate courses in design and disaster-risk management and, through exploring a case study of a design studio in Vietnam, illustrates the nature of the pedagogical principles involved.

2. Why Design?

Despite the growth of disaster studies as a field of research, especially in relation to the built environment (Sanderson, Kayden, and Leis 2016) [21], the disaster-management field is characterized by at least three weaknesses: (i) conceptual papers about the meaning and use of concepts with a lack of attention to operational issues; (ii) discrete case studies of disaster events and projects with little potential for meaningful generalizations; and (iii) single-sector and single-discipline studies that pay little attention to interdisciplinary processes, such as design, which can integrate across sectors and disciplines, and which are the key to addressing 'wicked problems'. A key consequence of these weaknesses is the lack of a systematic and analytical body of work that bridges the conceptual and operational gap between *knowing the causes and impacts of vulnerability and developing solutions* that ensure landscapes and settlement infrastructure are resilient and better capable of safeguarding communities vulnerable to disasters. Addressing this weakness helps answer a call from Alexander (2016) [22] for 'a major revision in the body of disaster theory' so that 'policies and practical solutions' can be derived from a conceptual road-map 'that clarifies complex realities and enables disasters to be managed and abated' (p. 2). The processes of design through systems analysis and design thinking offer a pathway to this new approach to theory and practice in disaster mitigation and recovery.

In relation to the built environment, design specifically refers to a problem-solving process based upon cross-disciplinary systems thinking, spatial innovation, and creativity. Design involves iterative sequences of observation to understand and empathize with the needs of people through their participation and collaboration, the rapid prototyping and testing of alternative socio-spatial designs (Boer et al., 2013) [23], design development, and full project implementation and evaluation. These are among the core skills of architects and other built environment and landscape designers. The development of such capabilities involves an on-going, iterative process of creating and reforming two- and three-dimensional space based on an understanding of the interdependence of human aspirations, the natural environment, and the social, political, and cultural systems in which the designs are constructed. Thus, design is 'both a mindset and a methodology' (Vandenbroeck 2012, p. 33) [24]. This integrative nature of design can help redress some of the challenges in successful disaster mitigation and recovery.

In what follows, 'design' and 'design thinking' are used with practical architectural and spatial applications in mind, encompassing the more transferable aspects of these terms in other disciplines, but which are particularly applicable to some of the great challenges in disaster mitigation and recovery. One of the greatest physical, financial, and social infrastructure losses after (un-)natural disasters is housing, which is a critical element in disaster mitigation and long-term community recovery. A home provides not only shelter from the elements, but also essential family stability, dignity, and security, especially after a disaster. However, there is a discrepancy between, on the one hand, the people-centered nature of architectural design (and 'housing as a process' as per Davis 1978, [25] 2015), and the product-delivery culture that characterizes many disaster-recovery programs. The result for housing is often a 'one size fits all' approach (Daly and Feener 2016) [26], inspired perhaps by a search for universal solutions for reasons of speed and economy. However, in this universal approach to rebuilding housing, insufficient attention is often paid to the aspirations of the people most affected and the infrastructure they specifically need, including the use of local housing technologies and site vulnerabilities. Tran (2018), for example, provides an illuminating analysis of the ways in which the "universalist' approach contributed to the failure of a disaster-resistant housing project in Vietnam. Architects' expertise in people-centered housing and settlement design is essential; yet, as Brett Moore (Chief, Shelter Division, UNHCR) [27] noted after Typhoon Yolanda in the Philippines: '*Despite the enormity of the disaster, it is almost impossible to get trained architects or planners for the complex task ahead of rebuilding the shelters and settlements*' (in Charlesworth 2014). [17] Nonetheless, approaches that use design to address the complex challenges of disaster mitigation and recovery are emerging through the intersection of research and field practice. Examples of such design-led responses, and an analysis of the lessons that may be learnt from them, are provided in the next section.

3. Disaster-Resilient Design

The frequent lack of systems thinking and design thinking in traditional risk-assessment and -reduction strategies was critiqued in the previous section. Nevertheless, they remain essential measures in disaster management. Essential, yes, but not sufficient, for as Geis (2000) [6] argued, 'truly safe' buildings and infrastructure are not possible 'without also having a safe overall community and region in which to build'. He went on to argue that '*[T]he only real way to minimize the growing human and property losses from earthquakes, cyclones, and severe flooding is rooted first and foremost in how we design and build our communities in the first place in . . . hazard-prone areas*' (p. 152). In integrating systems analysis and design thinking, Geis proposed the following elements as critical to what he termed 'disaster resistant design':

- An analysis of the overall capacity, functioning, and relationships of the various components and systems that support communities, business, and industry
- Integrating new development projects within the limits of natural systems
- The planning and design of development and redevelopment patterns
- The design and patterns of open space
- The design of neighborhood and commercial districts
- Individual and building group design, including location, configuration, and coherence with building code and climate-change imperatives
- The location, design, and service capacity of community facilities and public infrastructure
- Design to facilitate emergency management functions, including egress and access, the location, safety, and capacity of emergency shelters used and staging areas
- Utilizing maintenance and rehabilitation management as important tools for climate-change mitigation and adaptation (p. 157).

Similar sets of principles or frameworks have been proposed in the two decades since Geis proposed his list (e.g., Fisher 2013 [9], American Society of Landscape Architects 2020 [20], Lee 2020 [14]). Mostly, these have been couched in terms of disaster-resilient design, which has been defined as '*the intentional design of buildings, landscapes, communities, and regions in order to respond to natural and manmade [sic] disasters and disturbances—as well as long-term changes resulting from climate change—including sea level rise, increased frequency of heat waves, and regional drought*' (Wilson 2014) [28]. As the founder of the Resilient Design Institute, Wilson has also provided sets of principles and strategies for resilient design (see Resilient Design Institute, nd [29]).

These broad principles and strategies for resilient design were explored in a series of interviews with 15 architects working in disaster mitigation and recovery, and the analysis of their projects (Charlesworth 2014) [17] as well as in an analysis of the design principles behind exemplar design projects after disasters in Australia, Bangladesh, Haiti, Sri Lanka, the USA, and Vietnam (Charlesworth and Ahmed 2015) [12]. In addition, several core themes about design for disaster resilience, which transcend the practicalities of design and architecture, also emerged from this research. Evans (2015) [30] noted five of these as follows:

1. Architects have practical mind- and skill-sets which are of significant value in disaster mitigation and recovery, including the interdisciplinary understanding of science, engineering, technology, and materials; a spatial perspective on systems and patterns; creative problem-solving; planning, organizing, scheduling, and managing of—and working with—economic, social, emergency, legal, and governmental constraints.

2. The spatial awareness, aesthetic, and design skills that good architects bring to projects, the ability to create beauty—and perhaps in the most unlikely environments—do add real value to psychologically distressed and demoralized individuals and communities. This may not be a high priority in the immediate aftermath of a disaster, when the overwhelming need is simply to provide emergency shelter for thousands, perhaps hundreds of thousands, of displaced people. However, it is certainly very relevant in the transitional and permanent stages of rebuilding and resettlement.

3. The poor, marginalized, and the distressed deserve the benefits of good architecture equally, if not more so, than the privileged few who can afford the aesthetic and functional benefits of commercial design practices. As Shigeru Ban said in his acceptance speech for the Pritzker Prize in 2014, *'Architects are not building temporary housing because we are too busy building for the privileged people . . . . I'm not saying I'm against building monuments, but I'm thinking we can work more for the public'*.

4. There are no universal, one-size-fits-all solutions in resilient design. The most successful schemes, in terms of both their affordability and their benefits, are those built around intensive, sustained consultation with local people; the use, as far as humanly possible, of local materials and construction systems; and the employment of local people—often in situations where there is no other employment available—in the construction process.

5. Design education has not served the field of disaster-resilient design well. None of the 15 architects interviewed by Charlesworth (2014) [17] had encountered the concepts and practices of public-interest design (Adendroth and Bell 2019) [31], humanitarian architecture (Zuckerman Jacobson and Ban 2014) [32], or any related fields. Indeed, they often lamented that the kind of professional attitudes and ambitions that were encouraged during their training mitigated against a view of architecture as a community service akin to public health or human rights law in medical and legal education. Instead, they came to the field of design and disasters as the result of personal and family values and career aspirations to expand their sometimes-limited disciplinary backgrounds.

These five themes speak not only to the opportunities available to expand the scope and impact of architecture and other built environment professionals, creating a wider 'spatial agency' (Awan, Schneider and Till 2011) [33] for designers. They also provide a case for the reform of architectural and design education. As Till (2020) [34] noted, increasing the social engagement values and skills of designers requires a review of the aims, content focus, and culture of much architectural education, which, he argues:

> . . . inevitably edits down the social context of any project: rushed site visits, often abstract briefs with no clear user or client to engage with, and compressed timescales all mitigate against development of the skills required for socially engaged architecture. In addition, the standardized diet of juries, long nights, and isolation from other disciplines further consolidates the de-socialization of architecture students as they are admitted into the rituals of the tribe. A move towards a more socially engaged practice therefore needs a distinct shift in the processes, projects, and ethos of architectural education.

The next section reviews ways in which designers are increasingly being educated to work in disaster-resilient design, and includes a case study of a sample studio that illustrates the pedagogical principles appropriate to this.

4. Learning for Disaster-Resilient Design

The relative lack of attention to disaster management in design education has been noted for well over a decade. Writing in 2004, one commentator noted that this reflected the general neglect of design responses to national and global issues in the curricula of most architectural and design education programs (Bristol 2004) [35]. Papers by Griffith (2004) [36] Lloyd-Jones (2006) [37] and Lloyd-Jones et al. (2009) [38] Owen and Dumashie (2007), [39] Cage et al. (2009) [40] Wang (2010) [41] and Thurairajah et al. (2011) [42] confirmed this neglect and, along with Bristol, offered a range of content and skill objectives for integration into design education, specifically, and also into professional education in the wider built-environment disciplines. However, by 2016, Acar and Yalcinkaya (2016) [43] still noted that the body of literature on integrating disaster-management perspectives into the architecture curriculum remained 'relatively scarce' (p. 4), in that 'the number of undergraduate and post-graduate programs which integrate disaster-management perspectives into their curriculum as a long-term proactive strategy to build resilience is very low' (p. 1). Even in 2018, it was being lamented that the 'expansive pedagogic practices' envisioned for design education 'remain the exception rather than the norm', with many common practices, especially in architectural education, criticized for mitigating 'against development of the skills required for socially engaged architecture' (Till 2018, p. xxix) [44].

The situation at the master's degree level seems to be more amenable to innovation. For example, The Graduate School of Design at Harvard University is a key partner in the Harvard Humanitarian Initiative and has a master's degree specialization in Risk and Resilience. Two universities in Australia have established master's degrees in disaster-risk management within Schools of Architecture whilst universities in many countries have individual courses on architecture for climate-change adaptation and disaster resilience within their postgraduate degrees. In Europe, a small but growing group of university schools of design and architecture has been meeting since 2016 to explore and share developments and practices in specialist master's degrees on design for disaster and displacement. The objectives of these symposia included: building teaching and research collaborations; cross-enrolment of students in courses and field studies across the universities; and knowledge exchange with leading international delivery agencies, such as IFRC, UNHCR, and UN-Habitat. Participating universities include: Aalto and Hanken in Finland; UCL, Oxford Brookes, and Westminster in the UK; ETH in Switzerland; UIC in Spain; Paris-Belleville in France; KU Leuven in Belgium; University of Naples, Milan Polytechnic, and University of Venice in Italy; and RMIT from Australia. Each of the universities has written case studies of their master's programs and the pedagogical principles underpinning the curriculum, including reviews of particular design-education activities and studies. For example, KU Leuven provided cases studies of skill development in landscape urbanism to address flood problems in the Yangtse River Delta in China and the Guayas River Delta in Ecuador, while Westminster and RMIT reported on field studies on design responses for climate adaptation in Vietnam. (Reports of the case studies and discussions from the 2018 and 2019 meetings may be found under the heading 'International Symposia' at https://harbureau.org/#publications).

Drawing upon their experiences of teaching disaster-resilient design, this group has identified several recurring conceptual, ethical, and operational themes, including:

- The importance of critical reflection on design for disasters and displacement as the 'new normal'
- The desirability (or otherwise) of a competency framework for curriculum development in the field of design for disaster mitigation and recovery
- The practice of designing for a much wider range of clients than in commercial architecture, many of which are marginalized and may have few resources
- The ethical and political dimensions of design as a break with traditional 'modernist' practice in the profession
- The importance of teaching ethics in design education

- The challenge of teaching the values and skills underlying socially engaged co-design practices
- The value of integrating teaching and learning with research—co-produced, evidence-based practice
- Pedagogical challenges such as integrating conceptual knowledge of key issues through field-based studies and simulations.
- Integrating systems thinking and design thinking as pedagogical and professional tools.
- The value of developing transferable, 21st-century skills and predispositions suitable for employability in the disaster and humanitarian fields, especially for working with vulnerable people living under hardship.

5. Climate Change, Design, and Development Study, Hội An, Vietnam

Many of these issues can be illustrated through a case study of an international design studio conducted in Vietnam, which was presented at the European network's 2019 symposium. Convened by RMIT University, the studio was part of a semester-long course called 'Climate Change, Design and Development', which is a module in a degree called the Master of Disaster, Design and Development (MoDDD). The degree provides an early- to mid-career qualification for people wishing to transition their careers into the disaster-management field and, to cater for this, are taught through a blended learning format (online and face-to-face). The underlying philosophy throughout all courses is systemic design (an integration of systems and design thinking), which is used as a process for understanding issues and addressing problems in the complex and dynamic field of disaster-risk management. This means that the program is interdisciplinary and involves courses and students not just from architecture and planning but also engineering, communications, social science, project management, development studies, and environmental management. There is a strong emphasis on the promotion of research and operational skills for future employment through regular and intensive interaction with industry professionals through weekly webinars, field studios, intensive workshops, internships, and a capstone industry research project that students undertake as their final course. (Details of the degree program may be found at www.masterdisasterdesigndevelopment.com).

The learning objectives of the 'Climate Change, Design and Development' studio in Vietnam included developing skills for:

- Synthesizing knowledge from a variety of scientific and community-based sources on climate change, and the links between climate change and disasters
- Evaluating key strategies of climate-change adaptation and disaster-risk reduction, and their differences and convergences
- Interpreting and analyzing the implications of climate change and disasters for the built environment, in parts of the Asia–Pacific region, from diverse perspectives and sectoral linkages
- Working effectively with others in a field-based situation and demonstrating social, intercultural, and environmental awareness
- Communicating using diverse formats and strategies to engage with a range of stakeholders.

There were two parts to the course. Part 1 explored key concepts and trends in climate change and strategies for adaptation through a set of online materials, webinars, and workshop assignments. Part 2 was a design study on the design of adaptation strategies for the historic town of Hội An and its environs. This brief case study focuses on the design study. A cross- disciplinary group of ten architects, landscape architects, urban designers, planners, and built-environment professionals, as well as people with backgrounds in journalism and international development, participated. The study built upon the online studies in Part 1 and involved the sequence of activities in Table 1: (i) Site familiarization; (ii) Consultations and workshops with key local experts and stakeholders; (iii) Field investigations and data analysis; (iv) Design and planning exercises; and (v) Presentation and reporting.

Table 1. The sequence of learning activities and associated learning principles in the case study of learning to design for disaster resilience.

Phases in the Study	Learning Activities	Pedagogical Principles
1. Site familiarization	<ul><li>An annotated mapping and photographic exercise focusing on the biophysical landscape, the centrality of the river and flood plain to economic activities, and architecture and culture heritage</li><li>Statistical analysis of demographic and economic change in Hội An and surrounds</li></ul>	<ul><li>Field immersion</li><li>Intercultural communication</li><li>Stakeholder meetings</li><li>Systems analysis</li><li>Focus on socio-ecological relationships and drivers of change</li></ul>
2.Consultations and workshops with key local experts and stakeholders	<ul><li>Climate-change science and impacts in Vietnam</li><li>Flood-risk scenarios for Hội An region</li><li>Resilience Index research in Vietnam and Hội An</li><li>World Heritage values in Hội An and climate-change threats to, and impacts on, heritage values</li></ul>	<ul><li>Stakeholder meetings</li><li>Respect for scientific, social-science, and cultural knowledge and evidence</li><li>Synthesis of knowledge forms</li><li>Identifying design implications</li></ul>
3.Field investigations and data analysis	<ul><li>Situational Analysis</li><li>Coastal and wetland vulnerability analysis</li><li>Typhoon-resistant housing</li><li>Gender issues in climate-change adaptation</li></ul>	<ul><li>Vulnerability analysis and mapping</li><li>Recognition of traditional knowledges and flood adaptations</li><li>Design and precedent analysis</li><li>Gender sensitivity and design</li><li>Stakeholder consultation</li></ul>
4.Design and planning	<ul><li>Flood- and typhoon-resistant house designs</li><li>Landscape adaptations to flooding and coastal erosion</li><li>Design of an Adaptation Pathways Plan</li></ul>	<ul><li>Linking systems thinking and design thinking through systemic design processes</li><li>Development of design provocations</li><li>Stakeholder consultation and revision</li><li>Adaptation Pathways Planning</li></ul>
5. Presentation and reporting	<ul><li>Presentation preparation</li><li>Presentation to key stakeholders encountered in various activities</li><li>Report writing</li></ul>	<ul><li>Oral and written communication</li><li>Stakeholder consultation</li><li>Report writing and design</li><li>Podcast production</li></ul>

Figure 1 is a collage of images that illustrate this process. The primary objective of the study was to develop a number of design recommendations in Hội An to help build local capacity to adapt the built environment for climate resilience. Given the increasing pressures of tourism and sea-level rise on the town, planning ahead to mitigate such risks is critical for the future of the environmental, political, and cultural landscape of this central Vietnamese town.

1.

2.

Figure 1. *Cont.*

3.

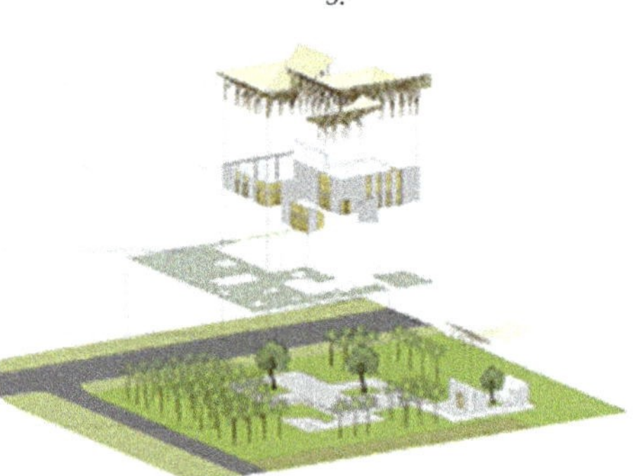

4.

5.

Figure 1. Images of the learning activities in the Hội An case study. **1.** Site familiarization: Group cycle trip to identify infrastructure vulnerabilities along the road from Hội An to the coast. **2.** Workshop by local experts: Heritage in Hội An. **3.** ((**a**) left) Field investigation: The use of wetlands to protect against storm surges during typhoons. ((**b**) right) Data Analysis: Vulnerability map-ping. **4.** Design: Planned green spaces around the city aim to slow down urbanization and limit further increases in building density, making way for more water runoff. **5.** Reporting: Interviewing householder for a podcast.

Climate-resilient designs for Hội An were developed along a transect from the Historic "Old Town" towards the coast and estuary. Figure 2 depicts some of the design-based recommendations that were developed for these two precincts. They illustrate the application of findings from the systems analysis that occurred in Phases 1–3. These included the continued use of two-story construction such as in the traditional adaptation techniques of shop-houses in the "Old Town", where light furnishings and retail stock from the ground floor can be raised to the second floor during flood periods. They also included the design of mangrove planting to provide a resilient nature-based solution to increased coastal erosion, and building typhoon-resistant houses on the urban fringe on stilts, rather than concrete slabs, as both flood and typhoon protection, and to ensure that the expansion of housing into the peri-urban padi-fields does not increase runoff by reducing the permeability of the soil to rain and flood water.

A summary of the report was presented to a workshop attended by twenty Hội An City and Quảng Nam Province officers and community representatives. These stakeholders recognized the value of the report but were also able to identify ways in which the study could be improved in the future. This included working even more closely with key stakeholder groups and focusing on adaptation needs in additional parts of the region. In addition, they offered to provide support to study additional issues in future design studies, e.g., quantitative data on mangrove-restoration schemes and construction density policies, building regulations, economic vs. human losses from flooding in Hội An, and social vulnerability. This advice is being integrated into future design studios for the course.

Old-Town Interventions

Figure 2. *Cont.*

Estuary Interventions

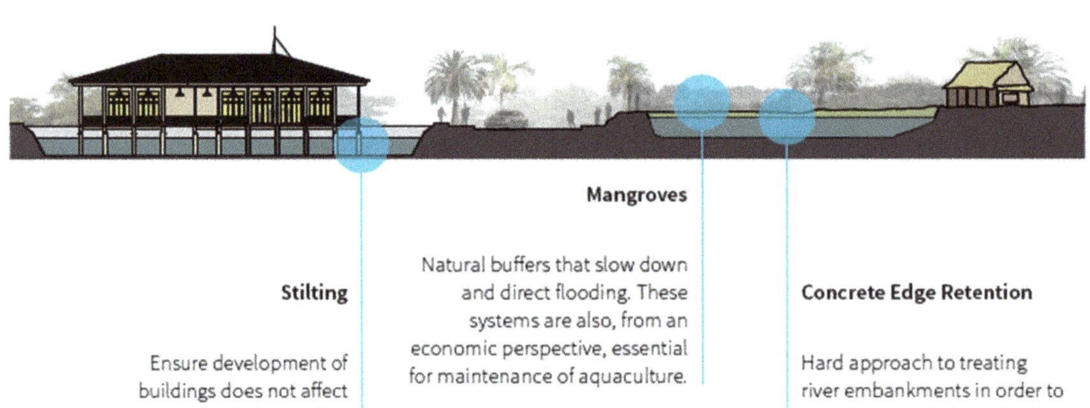

Figure 2. Recommended designs for climate-change adaptation in two precincts of Hội An.

Perhaps, the test of the efficacy of this study and the learning for disaster-resilient design that it sought to develop is to be seen in the vocational destinations of the students. Of the ten who undertook the 'Climate Change, Design and Development' course, four are now employed by development and disaster-based NGOs: Hannah by WADNA (Women and Development Network of Australia), Jaspreet by GAD Pod (Gender and Disaster Pod), Heidi by Live and Learn (from where she coordinated the COVID-19 communications program for the Australian government in the South Pacific), and Bjorn (a consultant for Oxfam on the design of blockchain for cash payments post-disaster, and now innovation Manager with the Humanitarian Innovation Fund). Nikhila is an Innovation Fellow in a School of Architecture in Australia; Carolina and Victoria are continuing postgraduate studies in international development; and Junyang, Alex, and Yingjie have returned to practice in landscape architecture with an increased commitment to disaster resilience in design.

6. Conclusions

This paper has sought to provide answers to the two questions posed in the introduction: (i) What can be learnt from examples of design-based responses to disaster mitigation and recovery? (ii) How can the skills needed to integrate design into disaster-risk management be developed? In doing so, it explored issues and lessons from design-based responses to disaster mitigation and recovery to develop a rationale for increasing the use of design strategies in addressing some of the 'wickedness' in disaster-risk management. It also provided principles and an example of how the skills needed to integrate design into disaster-resilience projects can be developed in future curricula in undergraduate and post-graduate design education and, thus, increase the interest and capabilities of design professionals in contributing to disaster-risk management.

In investigating the role of design in mitigating the damaging impacts of disasters and conflict, the paper has explained how design can be used as a strategic tool in developing interdisciplinary and innovative solutions to the often intractable, but ever-increasing, risks of (un-)natural disasters, sea-level rise, and human displacement. With unprecedented numbers of people globally rendered homeless by disasters, it is important to understand how previously under-utilized disciplines, such as architecture, can contribute to building disaster resilience. Other key learnings from the paper include:

- International, government, and community agencies are struggling to implement effective strategies for disaster-risk reduction and for planning long-term recovery after disasters. The central challenge these agencies face is the development of policies and practices for reducing the vulnerabilities that prevent communities from becoming more disaster-resilient. Design can be seen as a critical bridge in planning for disaster mitigation and recovery.
- There is a practice–theory gap between the community-led processes needed for long-term recovery and the product-delivery culture that characterizes many shelter and settlement programs. The result is often a 'one size fits all' approach to housing, with insufficient attention paid to the aspirations of the people most affected and the infrastructure needed.
- The skills of experienced system and design thinkers, such as architects, urban planners, and landscape architects), are seldom employed in the disaster-risk-management field, despite their demonstrated capacity to work with communities and to develop integrated spatial responses to guide both disaster-risk reduction and long-term rebuilding after a disaster. Developing design solutions at housing and settlement scales, e.g., preparing house designs and community master plans, is the core competency of architecture. However, this expertise has been neglected and the number of built-environment professionals such as architects equipped to respond in such situations is still very low.
- While there is an innate conservatism in most design degrees in terms of dealing with critical social challenges and crises, specialized masters degrees incorporating disaster-resilient design are emerging, and are training the next generation of disaster, humanitarian, and development professionals.
- The paper has outlined the contributions of design as a disciplinary and operational tool to deal with many of the social, environmental, and economic crises now being faced. However, a reorientation of design education is needed so that it addresses core disaster-risk-management concepts, such as vulnerability, urban resilience, climate-change adaptation, risk-based design, and scenario and community planning. Otherwise, it will not achieve its potential value in enhancing disaster resilience.

Author Contributions: Conceptualization, E.C. and J.F.; Writing—original draft, E.C. and J.F.; Writing—review & editing, E.C. and J.F. All authors have contributed equally to the paper and have read and agreed to the published version of the manuscript.

Funding: This research received no external funding.

Institutional Review Board Statement: Not applicable.

Informed Consent Statement: Participants agreed to photographs and graphics from the design being published.

Conflicts of Interest: The authors declare no conflict of interest.

References

1. Munich Re. Relevant Natural Catastrophe Loss Events Worldwide 2019. 2020. Available online: https://www.munichre.com/en/company/media-relations/media-information-and-corporate-news/media-information/2020/causing-billions-in-losses-dominate-nat-cat-picture-2019.html (accessed on 2 August 2020).
2. IDMC (Internal Disaster Monitoring Centre). *Global Report on Internal Displacement 2020*; IDMC, 2020; Available online: https://www.internal-displacement.org/global-report/grid2020/ (accessed on 2 August 2020).
3. IDMC (Internal Disaster Monitoring Centre). *Assessing the Impacts of Climate Change on Flood Displacement Risk*; IDMC, 2020; Available online: https://www.internal-displacement.org/publications/assessing-the-impacts-of-climate-change-on-flood-displacement-risk (accessed on 2 August 2020).

4. Cadman, E. *Insurance Industry Calls for Action to Mitigate Climate Risk as Australia Bushfires Widen*; Bloomberg News: New York, NY, USA, 2020; Available online: https://www.bloomberg.com/news/articles/2020-01-08/deadly-australia-fires-spur-calls-to-mitigate-disaster-risk (accessed on 9 February 2020).

5. Bojic, D.; Baas, S.; Wolf, J. *Governance Challenges for Disaster Risk Reduction and Climate Change Adaptation*; FAO: Rome, Italy, 2019.

6. Geis, D. By design: The disaster resistant and quality-of-life community. *Nat. Disaster Rev.* **2000**, *1*, 150–160. [CrossRef]

7. Aldunce, P.; Beilin, R.; Handmer, J.; Howden, M. Framing disaster resilience. The implications of the diverse conceptualisations of "bouncing back". *Disaster Prev. Manag.* **2014**, *23*, 252–270. [CrossRef]

8. Alexander, D.E. Resilience and disaster risk reduction: An etymological journey. *Nat. Hazards Earth Syst. Sci.* **2013**, *13*, 2707–2716. [CrossRef]

9. Fisher, T. *Designing to Avoid Disaster: The Nature of Fractal-Critical Design*; Routledge: New York, NY, USA, 2013.

10. Gonzalez-Mathiesen, C.; March, A. Establishing Design Principles for Wildfire Resilient Urban Planning. *Plan. Pract. Res.* **2018**, *33*, 97–119. [CrossRef]

11. Smith, T.; Choy, D.L. Adapting Australian coastal regions to climate change: A case study of South East Queensland. In *Climate Change and the Coast: Building Resilient Communities*; Glavovic, B., Kelly, M., Kay, R., Travers, A., Eds.; CRC Press: Boca Raton, FL, USA, 2015; pp. 269–284.

12. Charlesworth, E.; Ahmed, I. *Sustainable Housing Reconstruction: Designing Resilient Housing after Natural Disasters*; Routledge: London, UK, 2015.

13. Keenan, J. Seeking an interoperability of disaster resilience and transformative adaptation in humanitarian design. *Int. J. Disaster Resil. Built Environ.* **2018**, *9*, 145–152. [CrossRef]

14. Lee, A. *Design by Resilience*; Springer: Dordrecht, The Netherlands, 2020.

15. Buchanan, R. Wicked problems in design thinking. *Des. Issues* **1992**, *8*, 5–21. [CrossRef]

16. Johnson, T. 'Design Thinking' Is Changing the Way We Approach Problems. 13 January 2016. Available online: http://www.universityaffairs.ca/features/feature-article/design-thinking-changing-way-approach-problems (accessed on 12 February 2022).

17. Charlesworth, E. *Humanitarian Architecture: 15 Stories of Architects Working after Disasters*; Routledge: London, UK, 2014.

18. Elliot-Ortega, J. Urban Design as Problem Solving: Design Thinking in the Rebuild by Design Resiliency Competition. Master's Thesis, University of Chicago, Chicago, IL, USA, 2010. Available online: https://dspace.mit.edu/handle/1721.1/98931 (accessed on 26 July 2020).

19. Shaw, R.; Rahman, A.; Surjan, A.; Parvin, G. (Eds.) *Urban Disasters and Resilience in Asia*; Butterworth-Heinemann/Elsevier: Oxford, UK, 2016.

20. American Society of Landscape Architects. *Resilient Design*; American Society of Landscape Architects: Washington, DC, USA, 2020; Available online: https://www.asla.org/resilientdesign.aspx (accessed on 9 August 2020).

21. Sanderson, D.; Jayden, J.; Leis, L. (Eds.) *Urban Disaster Resilience*; Routledge: New York, NY, USA, 2016.

22. Alexander, D. The game changes: 'Disaster Prevention and Management' after a quarter of a century. *Disaster Prev. Manag. Int. J.* **2016**, *25*, 261–274. [CrossRef]

23. Boer, L.; Donovan, J.; Buur, J. Challenging industry conceptions with provotypes. *CoDesign* **2013**, *9*, 73–89. [CrossRef]

24. Vandenbroeck, P. *Working with Wicked Problem*; King Baudouin Foundation: Brussels, Belgium, 2012; Available online: https://www.kbs-frb.be/en/Virtual-Library/2012/303257 (accessed on 5 July 2015).

25. Davis, I. *Shelter after Disaster*, 2nd ed.; Oxford Polytechnic Press: Oxford, UK, 1978; Available online: https://www.ifrc.org/Global/Documents/Secretariat/201506/Shelter_After_Disaster_2nd_Edition.pdf (accessed on 18 November 2018).

26. Daly, P.; Feener, R. (Eds.) *Rebuilding Asia Following Natural Disasters: Approaches to Reconstruction in the Asia-Pacific Region*; Cambridge University Press: Cambridge, UK, 2016.

27. UNHCR. *Global Trends: Forced Displacement in 2019*; UNHCR: Geneva, Switzerland, 2020; Available online: https://www.unhcr.org/globaltrends2019/ (accessed on 2 August 2020).

28. Wilson, A. Resilience as Means of Mitigating Climate Change 2014. Available online: https://www.resilientdesign.org/resilience-as-means-of-mitigating-climate-change/ (accessed on 17 August 2020).

29. Resilient Design Institute (nd). What Is Resilience? Available online: https://www.resilientdesign.org/defining-resilient-design (accessed on 20 July 2020).

30. Evans, G. *Launch of Humanitarian Architecture*; Australian Institute of Architects, Victorian Chapter: Melbourne, Australia, 12 August 2014; Available online: https://www.gevans.org/speeches/speech549.html (accessed on 29 July 2020).

31. Adendroth, L.; Bell, B. *Public Interest Design Education Guidebook: Curricula, Strategies, and SEED Academic Case Studies*; Routledge: New York, NY, USA, 2019.

32. Zuckerman Jacobson, H.; Ban, S. *Shigeru Ban: Humanitarian Architecture*; Distributed Art Publishers: New York, NY, USA, 2014.

33. Awan, N.; Schneider, T.; Till, J. *Spatial Agency: Other Ways of Doing Architecture*; Routledge: London, UK, 2011.

34. Till, J. *Architecture after Architecture*; Routledge: London, UK, 2020; Available online: http://www.jeremytill.net/read/130/architecture-after-architecture (accessed on 2 August 2020).

35. Bristol, G. The last architect. In Proceedings of the Silpakorn Architectural Discourse 3rd Mini-Symposium, Bangkok, Thailand, 18–19 March 2004; Available online: https://www.academia.edu/7732516/The_Last_Architect (accessed on 9 July 2018).

36. Griffiths, R. Knowledge production and the research-teaching nexus: The case of the built environment disciplines. *Stud. High. Educ.* **2004**, *29*, 709–726. [CrossRef]

37. Lloyd-Jones, T. *Mind the Gap! Post-Disaster Reconstruction and the Transition from Humanitarian Relief*; RICS: London, UK, 2006; Available online: https://www.preventionweb.net/publications/view/9080 (accessed on 24 July 2020).
38. Lloyd-Jones, T.; Kalra, R.; Mulyawa, B.; Theis, M.; Wakely, P.; Payne, G.; Hal, N. *The Built Professions in Disaster Risk Reduction and Response: A Guide for Humanitarian Agencies*; MLC Press, University of Westminster: London, UK, 2009; Available online: https://www.ifrc.org/PageFiles/95743/B.a.07.Built%20Environment%20Professions%20in%20DRR%20and%20ResponseGuide%20for%20humanitarian%20agencies_DFDN%20and%20RICS.pdf (accessed on 24 July 2020).
39. Owen, D.; Dumashie, D. Built Environment Professional's Contribution to Major Disaster Management. In Proceedings of the FIG Working Week, Hong Kong, China, 13–17 May 2007; Available online: https://www.fig.net/resources/proceedings/fig_proceedings/fig2007/papers/ts_1h/ts01h_03_owen_dumashie_1531.pdf (accessed on 17 July 2020).
40. Cage, C.; Hingorani, D.; Jopling, S.; Parke, E. Building relevance: Post-disaster shelter and the role of the building professional. In *Proceedings of the Background Paper for Conference of the Centre for Development and Emergency Practice (CENDEP)*; Oxford Brookes University: Oxford, UK, 18 September 2009.
41. Wang, T. A new paradigm for design studio education. *Int. J. Art Des. Educ.* **2010**, *20*, 173–183. [CrossRef]
42. Thurairajah, N.; Palliyaguru, R.; Williams, A. Incorporating disaster management perspective into built environment undergraduate curriculum. In Proceedings of the International Conference on Building Resilience, Kandalama, Sri Lanka, 20–22 July 2011; Available online: https://www.researchgate.net/publication/333131988 (accessed on 31 July 2020).
43. Acar, E.; Yalçınkaya, F. Integrating disaster management perspective into architectural design education at undergraduate level—A case example from Turkey. In Proceedings of the 5th World Construction Symposium 2016, Colombo, Sri Lanka, 29–31 July 2016; pp. 284–293. Available online: https://www.researchgate.net/publication/303988576 (accessed on 28 March 2020).
44. Till, J. Foreword. In *The Routledge Companion to Architecture and Social Engagement*; Karim, F., Ed.; Routledge: London, UK, 2018; Volume xxvi–xxviii.

Review

Analysis of Renovation Works in Cappuccinelli Social Housing District in Trapani

Rossella Corrao [1,*], Erica La Placa [1], Enrico Genova [2] and Calogero Vinci [1]

[1] Department of Architecture, University of Palermo, 90128 Palermo, Italy; erica.laplaca@unipa.it (E.L.P.); calogero.vinci@unipa.it (C.V.)

[2] Energy Efficiency Unit Department (DUEE), Italian National Agency for New Technologies, Energy and Sustainable Economic Development (ENEA), 40129 Bologna, Italy; enrico.genova@gmail.com

* Correspondence: rossella.corrao@unipa.it

Abstract: The refurbishment of public residential districts represents a current and complex problem. The Cappuccinelli Social Housing (SH) district in Trapani, designed in the late 1950s by Michele Valori and built during the 1960s, is emblematic of the architectural quality and technological innovation of the time it was designed, but at the same time represents the physical and social decay that occurred just after its construction. The neighborhood was examined through a combination of inspections and documentary research. The inspections were conducted for the entire district in order to identify the recurrent external degradation of building components and the related causes, both physical and anthropogenic. This paper investigates the physical–mechanical degradation and problems connected to previous renovation work in this district. Furthermore, technological design solutions are discussed for deep renovation and energy efficiency improvement of one of the terraced buildings of the Cappuccinelli SH district.

Keywords: social housing; renovation works; energy efficiency improvement; Cappuccinelli Social Housing district; Trapani

Citation: Corrao, R.; La Placa, E.; Genova, E.; Vinci, C. Analysis of Renovation Works in Cappuccinelli Social Housing District in Trapani. *Architecture* **2022**, 2, 277–291. https://doi.org/10.3390/architecture2020016

Academic Editors: Iftekhar Ahmed, Masa Noguchi, David O'Brien, Chris Tucker and Mittul Vahanvati

Received: 15 February 2022
Accepted: 8 April 2022
Published: 12 April 2022

Publisher's Note: MDPI stays neutral with regard to jurisdictional claims in published maps and institutional affiliations.

1. Introduction

The importance of deep renovation to existing building stock, recently stressed by the European Commission [1], is well-established in Italian regulations and strategies on buildings and their performances [2].

Public housing districts are relevant in this context, as demonstrated by several researchers investigating their renovations in European countries [3]. Several efforts were dedicated to joining architectural retrofit aspects with energy-saving requirements. In [4], the energy renovation of a city district in a particularly challenging historical city center is presented. The authors carried out an energy assessment of different interventions applied to the SH district in Venice. The introduction of vegetation for climate mitigation in urban contexts and for the energy improvement of individual buildings is widely studied. In [5], the benefits arising from the application of nature-based solutions in urban districts are examined. A set of simulations showed an improvement in the overall comfort conditions of inhabitants in terms of "urban heat island" reduction and living wellness. In [6], the author argues that additions on the façade have considerable advantages for the recovery and improvement of the energy performance of buildings, and proposes this solution be applied to public housing buildings, in accordance with Gaspari J. in [7,8]. Interventions on the building envelope allow for not only improvements in the energy performance of buildings, but also improvements to their appearance and architectural quality, giving new life to public housing. BIM is a new method used to optimize the planning, construction, and management of building artifacts using the software. Through the use of BIM software (Last Planner, Plan Radar, Vienna, Austria), it is also possible to validate the economic benefits of the recovery of public housing. In this study [9], 4D BIM technology was applied to a case study in Northern Ireland to validate the effectiveness of the software in cases of social housing retrofit, a condition not particularly studied by the scientific community. The authors argue that their model could be expanded to all UK buildings. The authors of this article [10] analyzed Portuguese residential construction in order to demonstrate that the recovery of existing buildings, using a minimum economic investment and ensuring energy improvements, is essential for a sustainable economy. Roof renovation is the most cost-effective measure for improving heating and cooling thermal performance. A study conducted by researchers at the University of Seville [11] in Spain demonstrates how the application of a ventilated roof in a Mediterranean climate allows for a 65% reduction in cooling requirements without a consequent heating penalty.

Social Housing districts are a significant portion of the existing building stock and are frequently based on architectural solutions which are worth being preserved although not optimum for achieving the current technical requirements set for existing buildings. Furthermore, these districts are often affected by large amounts of physical degradation due to lack of maintenance and social marginalization. Finally, they often host vulnerable inhabitants, who should be the primary beneficiaries of public strategies for building renovation. In this regard, solutions aiming at improving both the buildings' energy efficiency and urban environmental quality are considerably effective and also provide relevant social outcomes [12].

The present paper focuses on the public housing district called "Cappuccinelli" in the Sicilian town of Trapani. Cappuccinelli is strategic in the sustainable development of Trapani since it is situated both along the waterfront and on one of the main roads of the town. At the same time, this area is vulnerable to hydrological risks since it is located on a swamp that was reclaimed at the end of the 1930s.

The district is the result of an interesting architectural project [13]. However, it also suffers from relevant physical decay (evident in the number of unauthorized additions and the lack of regular maintenance), which has fostered significant social problems. Because of these weaknesses, for a long time, the municipality ignored the potential of Cappuccinelli for its inhabitants and the town as a whole. Recently, thanks to an announcement issued as part of the National Innovation Program for the Quality of Living [14], the municipality has included the district among the sites selected for regenerating the urban, building, and socio-economic heritage of Trapani in order to improve social cohesion and guarantee the quality of life of citizens with a sustainable perspective.

This paper investigates Cappuccinelli as a reference case to develop planning and design procedures for the refurbishment of Italian public housing districts. In this sense, the circumstance that around 62% of dwellings in Cappuccinelli are owned by private parties—generally, the inhabitants themselves—and not by a public body is relevant given that this is a frequent barrier to urban-scale improvement strategies and renovations of Italian social housing districts.

2. Methodology

The research method adopted by the authors started with the analysis of Cappuccinelli Social Housing district through a detailed examination of a combination of documentary research, onsite inspections, and monitoring activities related to the thermal transmissions of external walls as discussed in previous work [15].

Documentary research was carried out at the archive of the public body "IACP Trapani", which is responsible for public dwellings in the Cappuccinelli district [13]. Two main objects were explored. First, project development was detailed from the awarded architectural concept to the construction of the district, which was characterized by several changes to the original design [13]. Secondly, technical documents of the renovation works carried out by IACP were examined and categorized according to the building and period of execution.

General onsite inspections were conducted for the entire district in order to individuate recurring outdoor degradations of the building components and their related causes, both physical and anthropogenic.

The alterations, classified according to the technical standard UNI 11182:2006 [16,17], were documented by means of photographs and geometric surveys. As far as the anthropogenic degradations are concerned, the types and locations of inappropriate external and structural changes, façade changes in façades (especially for openings), and unauthorized additions were reported systematically. Information collected about damages and alterations in the different blocks was related to the positions of the blocks in the district (especially for their exposure to wind and sea salt aerosol) and the rates of private and public owners per building.

As described in the following section, the study was widened to two blocks, where a detailed analysis of external degradation of the building envelope was carried out. For this purpose, a photogrammetric survey was carried out.

The additional data achieved through the focus on two blocks were used to define an order of priority for urgent renovation works [18] and discuss the technical proposals appropriate to match the deep renovation of the Cappuccinelli district with its architectural features.

3. Data Presentation Analysis and Discussion of Findings

3.1. Cappuccinelli Social Housing District

The "Cappuccinelli" district (Figure 1) was built as part of the national plan of social housing called "INA-Casa", which was implemented by Italian Law 43/1949. Architect Michele Valori and his group designed the district for a design contest in 1956.

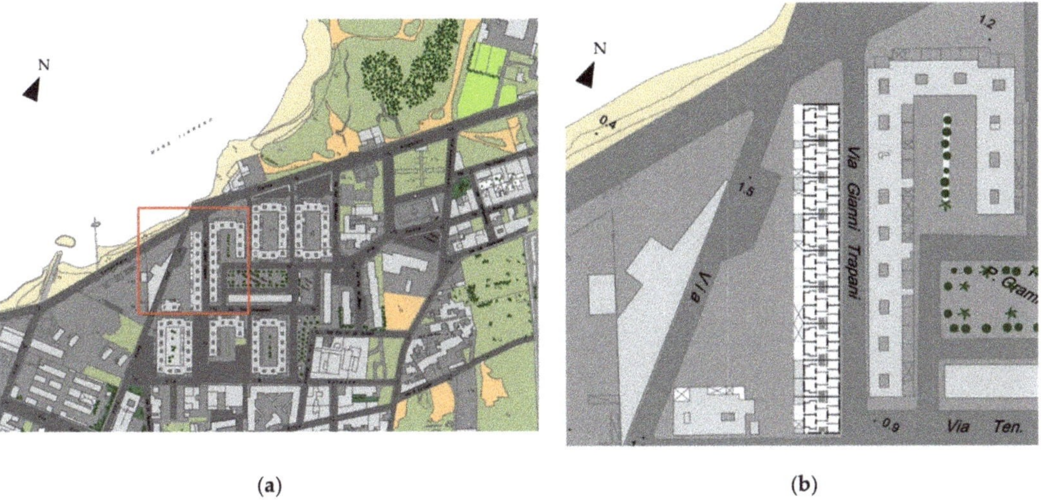

(a)   (b)

Figure 1. Plan of the Cappuccinelli District (**a**) and ground floor plan of the terraced block named "L" (**b**). Source: Rossella Corrao.

A large square is surrounded by two-storey blocks, each including large inner courtyards, and a multi-storey building. The idea, on which the project by Valori is based, creates a strict relation between private and public spaces that are linked by buildings, predominantly terraced houses with different typologies of apartments.

The "courtyard blocks" consist of duplex apartments with four to five rooms. These dwellings are organized in pairs and each one has a small courtyard that is accessible from a portico. This private courtyard is separated from the large common courtyard of the block, which is accessible from a secondary access point at the back of the apartment. The ground floor of each pair of apartments corresponds to a square. One side of this square is 12.60 m long and corresponds to the module that the district is based on. In the original project, the inner courtyard of each block was dedicated to extensive vegetation with private gardens, equipped areas, and a limited number of tall trees.

The phenomena of social decay impacted Cappuccinelli from the beginning. Many houses were illegally occupied even during the construction work, thus impeding the completion of the district according to the original project plan. Furthermore, unauthorized additions have proliferated: The small courtyards of paired duplexes were frequently closed and many private gardens in the large courtyard were changed to indoor spaces (Figure 2).

(**a**)   (**b**)

Figure 2. Unauthorized additions in the Cappuccinelli district: private gardens occupied by new rooms (**a**) and other chaotic modifications (**b**) in the terraced block named "L".

Physical and mechanical decay join the anthropogenic alterations of the courtyard blocks. Indeed, the buildings have a reinforced concrete-framed structure with infill walls comprising two layers of calcarenite blocks with an air cavity. Close to the sea and exposed to strong mistral winds, the structures suffer severe degradation from metal corrosion that has been accelerated by defective rainwater disposal from roofs. This implies extensive damage to the plasters and localized losses of the outer layer of the infill walls.

The analysis of the district has focused on the two blocks named "E" and "L" (Figure 3). Both have the same modular units as the courtyard blocks, namely a pair of duplex dwellings, but block "E" has an open courtyard while block "L" is a linear block of terrace houses.

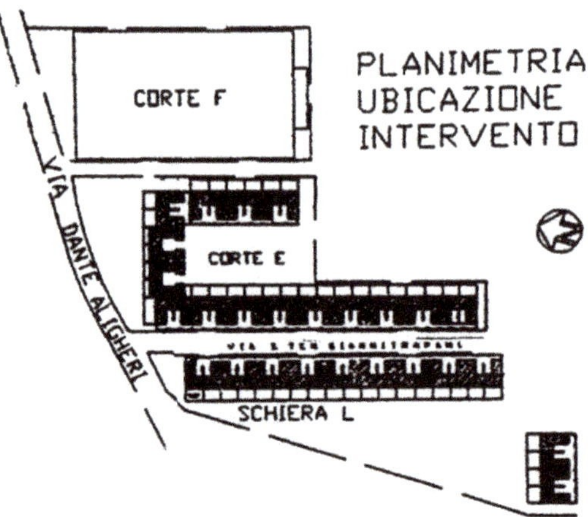

Figure 3. Historical site plan of "E" and "L" blocks. Source: Archive IACP Trapani.

3.2. Deep Renovation Works

In the examined blocks "E" and "L", renovation works have been carried out frequently in the last three decades, as reported in the archival documents [19]. The realized works include the demolition and reconstruction of the roofs, the restoration of reinforced-concrete structural elements, and the removal and replacement of plasters. Table 1 summarizes the works carried out from 1992 to 2005 in the two analyzed blocks.

Table 1. Renovation works carried out in blocks "E" and "L" of the Cappuccinelli district in the period from 1992–2005. Data source: Archive IACP Trapani.

Year	Block	Number of Dwellings	Renovation Works
1992	E	30	Structural refurbishment. Demolition and reconstruction of some floors. Replacement of plaster, downspouts, roof waterproofing, and toilets.
	L	22	
2000	L	1	Refurbishment of pillars and edge beams. Demolition and reconstruction of some floors. Replacement of external doors and windows. Roof waterproofing. Replacement of external infill walls on the first floor by using lightweight concrete blocks.
2001	E	3	Partial replacement of first-floor infill wall with 12 cm-thick perforated bricks. Refurbishment of pillars and edge beams. Construction of new roof slabs. Roof waterproofing. Replacement of external plasters on the ground floor. Decommissioning of products containing asbestos. Replacement of house water and electrical systems.
	L	1	
2003	E, L	3	Refurbishment of reinforced-concrete pillars and beams, and floors. Demolition and reconstruction of roof slabs with precast elements in brick concrete, 20 cm thick (16 + 4). Partial replacement of infill walls with precast concrete blocks subject to water repellent treatment. Dismantling of water tanks placed on the roof. Refurbishment of water and electrical systems in a dwelling.
2005	E, L	24	Refurbishment of reinforced-concrete structural elements. Demolition and reconstruction of roof slabs. Removal of water tanks from the roof. Replacement of severely oxidized external iron with windows consisting of PVC frames and double glazing. Replacement of external plaster and indoor wall finish. Technical improvement of electrical systems.

The design documents of these renovation works show that no particular attention was paid to the construction and aesthetic details of the original project by Valori. Furthermore, the technical solutions used in the first renovations were replicated without update or revision in the following years, for about one decade, despite the innovative materials and techniques introduced in the building market. Starting in 2001, IACP decided to enlarge the openings on the first floor of the main façades with porticos. This decision aimed to satisfy the necessities of inhabitants, namely increasing the natural lighting and ventilation of the indoor spaces. The architectural impact on the original project, namely the alteration of the geometric composition of the façades, was relevant but neglected, showing that the architectural quality of the project by Valori was not fully recognized.

3.3. Recurring Degradations in Cappuccinelli Buildings

Although several works have been updated over time, the current state of the buildings in Cappuccinelli is generally characterized by advanced decay. Many houses are even precarious, as shown by the recent collapse of a portion of the cornice in the portico of block "E".

Appropriate renovation works must be based on a more in-depth analysis of the causes of decay, which can be referred to as a list of recurring phenomena. An analysis of recurring degradations and their related causes was carried out according to what was suggested in [18]. Actually, Nowogońska B. and Mielczarek M. have determined a method for quantification of the cause–effect combinations of deterioration in order to define an order of priority for urgent renovation works and in this sense, the authors have quantified the decay of different Cappuccinelli building elements, focusing on building "L". This is the only block where an original dwelling designed by Valori still exists. The block is also very close to the sea and its façades are particularly exposed to the winds, as compared to the other buildings in the district. Furthermore, renovation works carried out in block "L" are largely documented in the IACP archive. Tables 2 and 3 describe the types of degradation observed in block "L" according to the technical standard UNI 11182:2006 [16,17,20].

Table 2. Degradations observed in the building elements of "L" block.

Degradations Typologies	Description	Cause	Photographic Reproduction	Affected Building Element
Staining	Localized chromatic alteration on the surface	Oxidation of metal elements		Façade
Plaster Exfoliation	Degradation that manifests itself with detachment, often followed by falling, of one or more subparallel surface layers (sheets)	Movement of water within the substrate; action of microorganisms		Ceiling
Plaster Bursting	Local loss of the plaster surface from internal pressure, usually in the form of an irregularly sided crater	Differential dilatations between support and finishing materials		Façade
Plaster Disintegration	Decohesion characterized by detachment of aggregates under minimal mechanical stress	Infiltration of water; reaction between building materials and atmosphere		Façade
Plaster Erosion	Loss of the original surface, which leads to discontinuity of the plaster	Mechanical erosion from driving rain; capillary rise; erosion by abrasion of the cortical layers caused by the wind		Façade
Plaster missing layer	Solution of continuity between surface layers of the material, both between them and with respect to the substrate: generally, a prelude to the fall of the layers themselves	Consistent presence of saline formations; efflorescence; continuity solutions resulting from the presence of cracks and structural lesions; errors in installation and use of unsuitable sands/mortars		Façade

Table 2. *Cont.*

Degradations Typologies	Description	Cause	Photographic Reproduction	Affected Building Element
Masonry Disintegration	Decohesion characterized by detachment of granules or crystals under minimal mechanical stress	Infiltration of water, capillary rise; reaction between building materials and atmosphere		Façade
Masonry Erosion	Removal of material from the surface due to different processes: erosion by abrasion (mechanical causes), erosion by wear (anthropogenic causes)	Mechanical erosion from driving rain; erosion by abrasion of the cortical layers caused by the wind		Façade
Hole	Fall and loss of portions of masonry	Localized leaks from waste disposal and/or water conveyance systems; consistent presence of saline formations; continuity solutions resulting from the presence of cracks and/or structural lesions; installation errors and the use of unsuitable sands/mortars		Façade
Crack	Individual fissure, clearly visible to the naked eye, resulting from separation of one part from another	Disruption of the supporting masonry system; physical–mechanical incompatibility between substrate and finish; differential dilatations between support and finishing materials		Façade
Efflorescence	Formation of substances, generally whitish in color and crystalline, powdery, or filamentous, on the surface of the plaster	Capillary rising damp; condensation; localized system losses; runoff of rainwater; presence of sulphates; wind action that accelerates the surface evaporation of water		Façade
Flaking	Degradation that manifests itself in the total or partial detachment of parts (flakes), often corresponding to the continuity solutions of the original material	Exposure to atmospheric agents; presence of humidity in the masonry (crystallization of soluble salts)		Façade

Table 3. Degradations of the reinforced concrete elements in the North-East elevation of "L" block.

Degradation Typologies	Description	Cause	Photographic Reproduction	Affected Structural Element
Outcropping armor	Reduced concrete cover, non-existent or expelled due to degradation in progress	Insufficient concrete cover and incorrect positioning of the spacers for the reinforcement		Beams, pillars, floor joists
Corroded armor	Formation of hydrated iron oxide (rust) on the surface of the reinforcement which causes the reduction of the resistant section	Infiltration of rainwater; incorrect design in construction details; use of marine sands		Beams, pillars, floor joists
Lowering of pH	Cancellation of the basic properties of concrete (pH$\simeq$13)	Presence of humidity and carbon dioxide; insufficient concrete cover; absence of protective coating of the concrete; porosity and permeability		
Structural subsidence	Changing the geometry of the element	Increase of the acting loads; repeated stress from loading and unloading cycles; absence of the transversal elements of distribution in the floor		Floor joists
Detachment of concrete cover	Separation of the surface concrete layer of the reinforcements it covers	Incorrect positioning of the spacers for the reinforcement		Beams
Crack	Long and deep cracks in the continuity of the concrete	Increase of the acting loads; previous interventions carried out incorrectly; inadequate structural dimensioning		Pillars
Rust spots	Reddish spots on the external surface of the concrete corresponding to corroded reinforcements	Interventions previously performed incorrectly		Façade
Breakage of steel	Total loss of traction resistance, fracture of the entire section of iron	Increase of the acting loads; repeated mechanical stresses; inadequate structural sizing; erroneous evaluation of overloads		Floor joists

A public body (IACP) currently owns seven of the 18 dwellings in block "L" (Figure 4). The only apartment not altered from the project by Valori is owned by a private party and suffers from more severe degradation than the other apartments in the building. Both the geometric and material features of this dwelling have been analyzed.

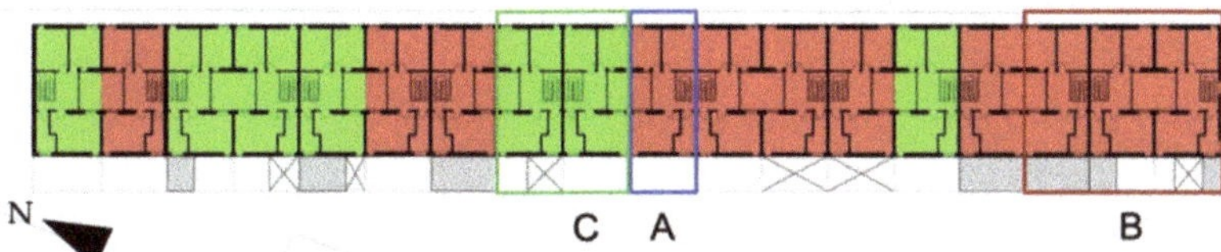

Figure 4. Ground floor of "L" block. Dwellings still owned by IACP are shown in green and private dwellings are shown in red. The blue rectangle identifies the apartment where the project by Valori has not been altered. Source: Archive IACP Trapani.

Figure 5a represents portion A (Figure 4) of the original plan (the project by Valori) and the southwest elevation of the "L" block and identifies its degradations: plaster exfoliation, plaster erosion, lack of plaster portions, masonry disintegration, masonry erosion, and lack of masonry blocks. In Figure 5b, the extent of each degradation phenomenon is estimated as a percentage of the analyzed surface. Plaster erosion is prevalent (31.9%), while other degradations range from 0.6 to 5.6%.

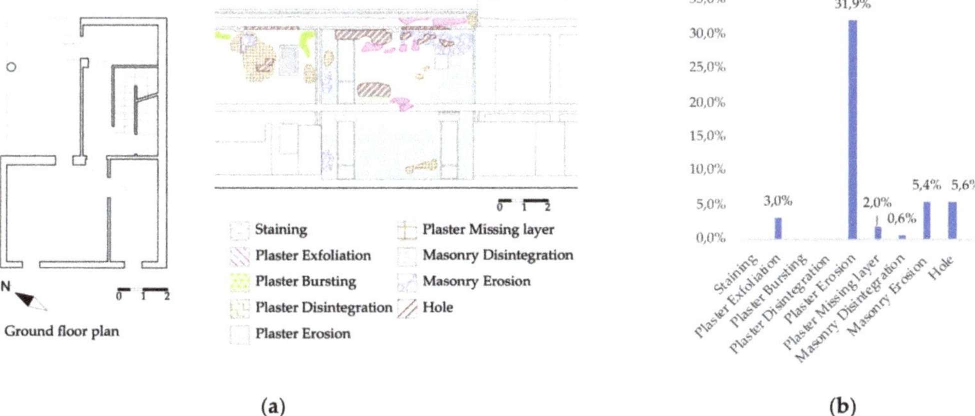

(a)   (b)

Figure 5. Ground floor plan from Valori project and representation of external visible damages on the façade (**a**), their extent estimated as a percentage of the analyzed façade surface (**b**) in the portion A (Figure 4) of the SW elevation of the "L" block. Authors: Erica La Placa, Calogero Vinci.

By extending this analysis to the entire "L" block (Figure 6a), erosion is confirmed as the most frequent degradation (4.5%) in all dwellings whose ownership passed from the public subject (IACP) to the private (original renters) (Figure 6b).

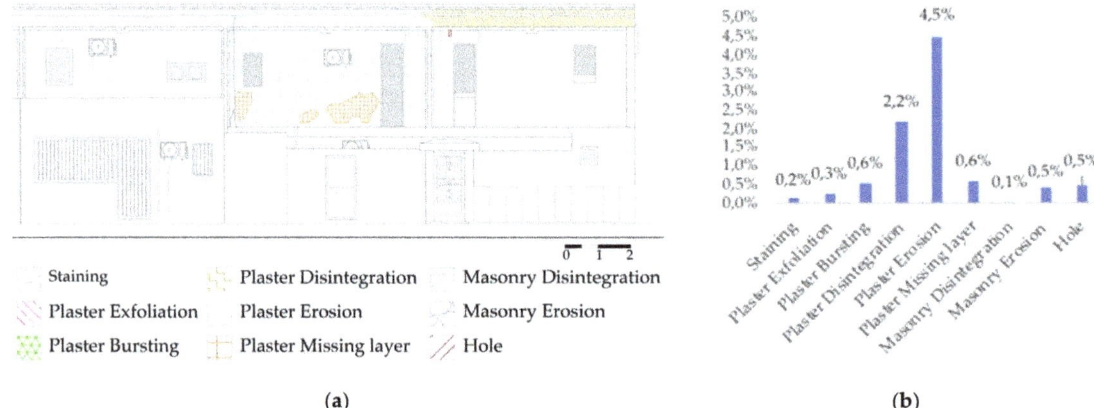

Staining   Plaster Disintegration   Masonry Disintegration

Plaster Exfoliation   Plaster Erosion   Masonry Erosion

Plaster Bursting   Plaster Missing layer   Hole

(**a**)   (**b**)

Figure 6. (**a**) Degradations of the portion B of the S-W elevation of "L" block in private dwellings. (**b**) Estimation of degradations as percentage of the analyzed surface. Authors: Erica La Placa, Calogero Vinci.

If the part still owned by IACP is considered (Figure 7a), the most recurrent deterioration is the lack of plaster portions (1.1%) (Figure 7b). More generally, the most frequent degradations on the façades of "L" block involve plasters. Indeed, the materials used for this functional layer were not suitable to calcarenite. Being very porous, this stone easily absorbs marine moisture and its content of salts, accelerating the mechanisms of the observed degradations.

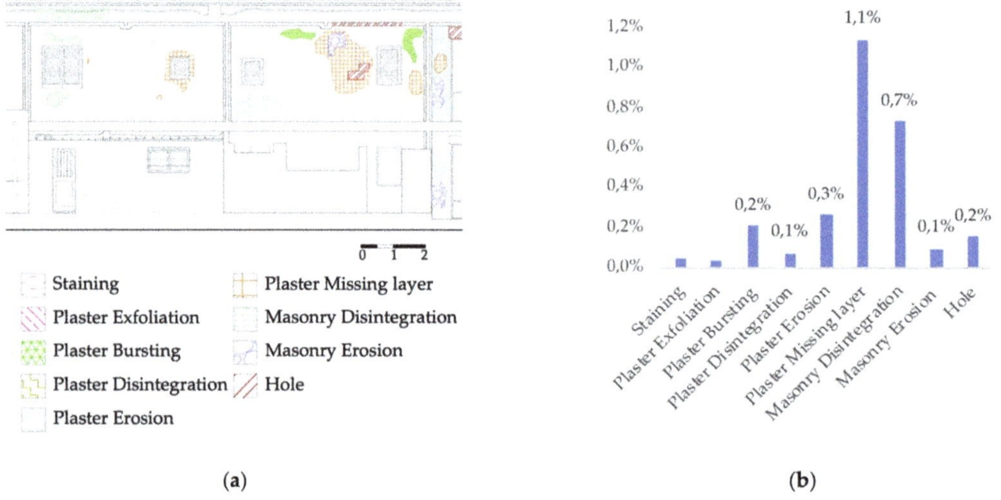

Staining   Plaster Missing layer

Plaster Exfoliation   Masonry Disintegration

Plaster Bursting   Masonry Erosion

Plaster Disintegration   Hole

Plaster Erosion

(**a**)   (**b**)

Figure 7. (**a**) Degradations of the portion C in the SW elevation of "L" block in dwellings still owned by IACP. (**b**) Estimation of degradations as a percentage of the analyzed surface. Authors: Erica La Placa, Calogero Vinci.

The degradations observed in block "L" are recurrent in Cappuccinelli and the information collected through this focus has been used to identify appropriate technological upgrades that could be implemented in large parts of the district both by private and public owners.

4. Suggestions for Appropriate Deep Renovation

Buildings in the Cappuccinelli district, being close to the sea (about 150 m) and exposed to strong mistral winds, should be equipped with protection systems for reinforced concrete structures in order to significantly reduce the risk of corrosion of the steel reinforcements. However, repairs by the IACP modified the original shape of these building components. As a relevant example, the original shape of the pillars in the porticos (Figure 8a)—a significant architectural detail in the project by Valori—was heavily modified in the renovation works, but their alteration did not make them more effective (Figure 8b).

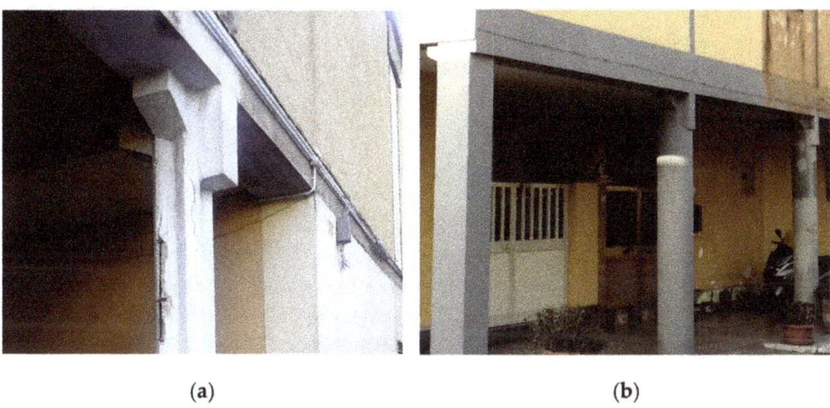

(a)   (b)

Figure 8. (a) Original shape of pillars designed by Valori. (b) Consolidation of pillars in the portico of court "B", executed in 2012. Author: Rossella Corrao.

The design of structural strengthening and rehabilitation works, aimed at overcoming the detected causes of damage (presence of rainwater and sea salts; lack of adequate protection for the reinforcement bars; inadequate structural dimensioning) and complying with current requirements for structural safety, also must consider the architectural concept on which the buildings are based. In this sense, archive research on project development and renovation works carried out in Cappuccinelli were essential to complement the in-situ inspection of the buildings in order to identify the original solutions worth being preserved and the inappropriate repairs.

The use of ventilated façades with plaster finishing and/or innovative living walls, discussed in previous works [9,10], can be considered effective solutions for the protection of the façades from the action of atmospheric agents that are particularly aggressive in the environmental context of Cappuccinelli. Their effectiveness has been examined both from environmental and energy points of view.

Ventilated façades prevent atmospheric agents (rain, wind) from damaging the structure, guaranteeing additional protection for the building and limiting metal corrosion, which widely affects the structures of buildings in the district.

Furthermore, if the ventilated façade is realized with plaster finishing (Figure 9), heat flow can be reduced up to about 70% due to the effects of the ventilation [21]. Compared to modular panels, an outer layer made of plaster minimizes the visual impact of the renovation and preserves the urban and architectural features of the buildings. Reduction in surface temperature, a decrease in solar radiation in summer, and the consequent improvement of indoor comfort (mitigation of overheating) are some of the benefits achieved through living walls (Figure 10), which contribute also to noise insulation and air purification. From this point of view, it is worth remarking that the extensive use of vegetation already characterized the original project by Valori, namely by private gardens and trees located in the main inner courtyards. The introduction of vegetation in order to improve energy performance (Figure 11) has been extensively discussed in previous works [15].

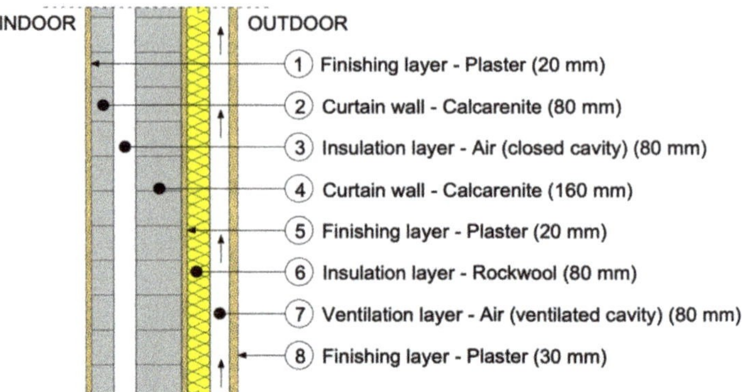

Figure 9. Ventilated façade with plaster finishing layer. Authors: Rossella Corrao, Erica La Placa.

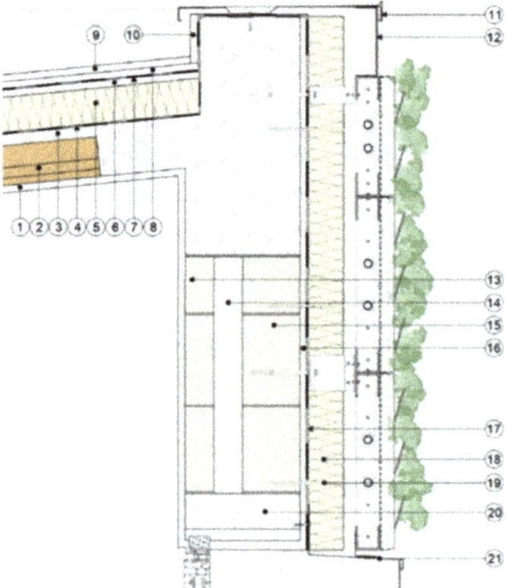

Figure 10. Detail of the living wall system tested to improve energy efficiency of the building envelope of the Cappuccinelli buildings. Authors: Mattia Bruno, Gaetano Di Fede, Giordana Motta. (1) Finishing layer: internal plaster, th. 2 cm; (2) Support element: brick block, h 12 cm; (3) Load distribution layer: concrete, 3 cm thick; (4) Vapor barrier layer: synthetic screen, th. 4 mm; (5) Thermal insulation layer: rigid rock wool panel, th. 10 cm; (6) Stiffening layer: concrete, sp. 3 cm; (7) Water tightness layer: double layer bituminous sheath, th. 4 mm; (8) Connection layer: cement mortar, sp. 2 cm; (9) Finishing layer: brick tiles, th. 2 cm; (10) Protection and finishing element: marble slab, th. 2 cm; (11) Water sealing element: shaped aluminum sheet, th. 2 mm; (12) Stiffening and support element: aluminum profile sp. 2 mm; (13) Resistant layer: block of calcarenite, sp. 8 cm; (14) Insulating layer: closed cavity, 8 cm; (15) Resistant layer: block of calcarenite, sp. 16 cm; (16) External finishing layer: plaster, th. 2 cm; (17) Vapor barrier layer: synthetic screen th. 2 mm; (18) Thermal insulation layer: rock wool panel sp. 10 cm; (19) Fastening element: nail with washer; (20) Load-bearing element: architrave in reinforced concrete, h 15 cm; (21) Water sealing element: shaped aluminum sheet, th. 12/10.

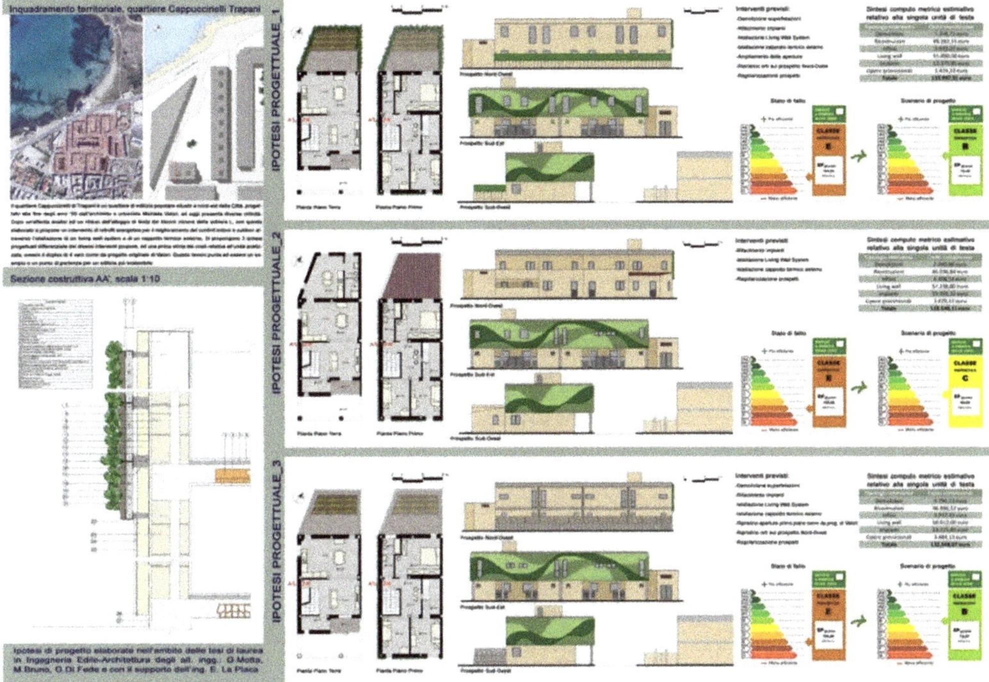

Figure 11. Summary table of the living wall installation on the façades of block "L". Authors: Mattia Bruno, Gaetano Di Fede, Giordana Motta.

5. Conclusions

The planning and design of effective building renovations should be addressed through a detailed examination of the physical–mechanical degradations and their causes, together with an analysis of technical and architectural features of buildings. This approach is necessary in the maintenance or refurbishment of social housing districts as well. This stock is often based on interesting architectural and technical solutions that are rarely recognized or frequently altered by both inhabitants and responsible public authorities.

In the Cappuccinelli Social Housing district, inadequate renovation works have merged with a lack of maintenance, social difficulties, and hydrologic vulnerability. On the other side, the architectural quality of the original project and the strategic urban location of this district could be a sensible basis to improve its environmental and social contexts.

The relevance of damages has been inspected also in relation to the ratio between public and private ownership in each block. The main difficulty in standardizing the recovery interventions is linked to the high percentage of apartments sold; in fact, 62% of the apartments are privately owned, while 32% are owned by the IACP. In order to solve this problem, the public body should propose design solutions that are suitable for the energy and architectural improvement of the buildings in close synergy with private individuals in order to uniformly redevelop the entire Cappuccinelli district.

In this study, the physical degradations and anthropogenic alterations of the buildings have been inspected and detailed for two blocks of the district. In addition, the relevance of the damage has been inspected in relation to the ratio between public and private ownership in each block.

The results suggest the relevance of systematic detection and mapping of degradations and alterations in the entire district. Indeed, this information is mainly based on visual, external observation and is able to overcome the difficulties of fragmented ownership.

On the one side, it would be useful to estimate the current structural and energy performances of a dwelling in the district, showing the most critical upgrade measures to implement and highlighting a priority order for urgent renovations. On the other side, by showing recurrent building deficiencies, the systematic analysis and categorization of damages support the identification of effective and replicable solutions to renovate the district buildings and to improve the quality of the urban space.

Author Contributions: Conceptualization, E.L.P.; methodology, E.L.P. and C.V.; software, E.L.P. and C.V.; resources, R.C. and E.L.P.; writing—original draft preparation, E.L.P., E.G. and C.V.; writing—review and editing, E.G. and C.V.; supervision, R.C.; All authors have read and agreed to the published version of the manuscript.

Funding: This research received no external funding.

Institutional Review Board Statement: Not applicable.

Informed Consent Statement: Not applicable.

Data Availability Statement: Data supporting reported results are available from the corresponding author on a reasonable request with the permission of IACP Trapani.

Conflicts of Interest: The authors declare no conflict of interest.

References

1. European Commission. *A Renovation Wave for Europe—Greening Our Buildings, Creating Jobs, Improving Lives*; 662-Final; European Commision: Brussels, Belgium, 2020.
2. Ministero dello Sviluppo Economico; Ministero dell'Ambiente e della Tutela del Territorio e del Mare; Ministero delle Infrastrutture e dei Trasporti. *Piano Nazionale Integrato per l'Energia e il Clima*; Ministero dello Sviluppo Economico: Rome, Italy, 2019.
3. Scanlon, K.; Fernández Arrigoitia, M.; Whitehead, C. Social Housing in Europe. In *European Policy Analysis (17)*; Swedish Institute for European Policy Studies: Stockholm, Sweden, 2015; pp. 1–12.
4. Teso, L.; Carnieletto, L.; Sun, K.; Zhang, W.; Gasparella, A.; Roma noni, P.; Zarrella, A.; Hong, T. Large scale energy analysis and renovation strategies for social housing in the historic city of Venice. *Sustain. Energy Technol. Assess.* **2022**, *52*, 102041. [CrossRef]
5. Mosca, F.; Dotti Sani, G.M.; Giacchetta, A.; Perini, K. Nature-Based Solutions: Thermal Comfort Improvement and Psychological Wellbeing, a Case Study in Genoa, Italy. *Sustainability* **2021**, *13*, 11638. [CrossRef]
6. Barbara, G. Optimization & Comparative Measures for the Strategy of Addition in Social Housing Retrofit in Italy in the 80's. In *Edilizia Circolare*; Cuboni, F., Desougus, G., Quaquero, E., Eds.; EdicomEdizioni: Monfalcone, Italy, 2018; pp. 949–958.
7. Gaspari, J. *Trasformare l'Involucro: La Strategia dell'Addizione nel Progetto di Recupero*; EdicomEdizioni: Monfalcone, Italy, 2014.
8. Gaspari, J. La strategia di addizione nei processi di riqualificazione energetica del costruito. In *Il Progetto Sostenibile*; EdicomEdizioni: Monfalcone, Italy, 2011; pp. 68–71.
9. Kemmer, S.; Biotto, C.; Chaves, C.; Koskela, L.; Tzortzopoulos, P. Implementing last planner in the context of social housing retrofit. In Proceedings of the 24th Annual Conference of the International Group for Lean Construction, Boston, MA, USA, 20–22 July 2016; pp. 83–92.
10. Palma, P.; Gouveia, J.P.; Barbosa, R. How much will it cost? *An Energy Renovation Analysis for the Portuguese Dwelling Stock. Sustain. Cities Soc.* **2021**, *78*, 103607. [CrossRef]
11. Delgado, M.G.; Ramos, J.S.; Amores, T.R.P.; Medina, D.C.; Domínguez, S. Álvarez Improving habitability in social housing through passive cooling: A case study in Mengíbar (Jaén, Spain). *Sustain. Cities Soc.* **2021**, *78*, 103642. [CrossRef]
12. Figliola, A.; del Brocco, B.; de Matteis, M. *Rigenerare la Città: Il Social Housing Come Opportunità di Rinnovo Urbano e Sociale*; IUAV: Venezia, Italy, 2014.
13. Corrao, R. Conoscere per valorizzare e rigenerare: Il progetto di Michele Valori per il quartiere Cappuccinelli a Trapani. In *Patrimonio in Divenire, Conoscere, Valorizzare*; ReUSO Matera: Rome, Italy, 2019; p. 1451.
14. Ministro delle Infrastrutture e dei Trasporti; Ministro dell'Economia e delle Finanze; Ministro per i Beni e le Attività Culturali e per il Turismo. Decreto Interministeriale n. 395. *Gazzetta Ufficiale della Repubblica Italiana*, 16 November 2020; n. 285.
15. Corrao, R.; Vinci, C.; Genova, E.; la Placa, E. Ipotesi di riqualificazione per un alloggio duplex del quartiere Cappuccinelli a Trapani. In *Design and Construction: Tradition and Innovation in the Practice of Architecture*; Sicignano, E., Ed.; EdicomEdizioni: Monfalcone, Italy, 2021; pp. 1309–1323.
16. UNI—Italian Organization for Standardization. *UNI 11182 Cultural Heritage: Natural and Artificial Stone: Description of the Alteration: Terminology and Definition*; UNI—Italian Organization for Standardization: Rome, Italy, 2006.
17. ICOMOS-ISCS. Illustrated Glossary on Stone Deterioration Patterns. Available online: https://www.icomos.org/publications/monuments_and_sites/15/pdf/Monuments_and_Sites_15_ISCS_Glossary_Stone.pdf (accessed on 28 January 2022).
18. Nowogonska, B.; Mielczarek, M. Renovation management method in neglected buildings. *Sustainability* **2021**, *13*, 929. [CrossRef]

19. Archive of Istituto Autonomo Case Popolari (IACP). *Trapani, Folder Rione Cappuccinelli Corte E—Schiera L*; Archive of Istituto Autonomo Case Popolari (IACP): Pistoia, Italy, 2021.
20. Corrao, R. *Patologie di Degrado del Cemento Armato*; Ila Palma: Palermo, Italy, 2004.
21. Corrao, R.; la Placa, E. Plaster ventilated façade system for renovating modern and ancient buildings. A CFD analysis. *IOP Conf. Ser. Earth Environ. Sci.* **2021**, *863*, 012046. [CrossRef]

Article

Significance of Occupant Behaviour on the Energy Performance Gap in Residential Buildings

Claire Far *, Iftekhar Ahmed and Jamie Mackee

Faculty of Engineering and Built Environment, School of Architecture and Built Environment, University of Newcastle, Callaghan, NSW 2308, Australia; ifte.ahmed@newcastle.edu.au (I.A.); jamie.mackee@newcastle.edu.au (J.M.)
* Correspondence: claire.far@uon.edu.au

Abstract: Buildings are an important part of worldwide efforts to reduce energy consumption and mitigate greenhouse gas emissions that contribute to climate change. Despite recent technological developments in the area of energy consumption reduction, energy use is on the rise, highlighting the significance of considering occupant behavior with regard to controlling energy consumption and supporting climate resilience. Energy performance of residential buildings is a function of various aspects such as properties of the building envelope, climatic location characteristics, HVAC system, and, more importantly, occupant behavior and activities towards energy utilization. This study carries out a comprehensive review of the impact of occupant behavior on reducing the energy performance gap in residential buildings since residential buildings account for 70% of building floor area around the globe. Findings have revealed that a dearth of literature on occupants' behavior scholarship leads to inaccurate simplifications in building modeling and design. Thus, there is a strong need to obtain appropriate occupant behavioral data to develop strategies to close the energy performance gap as much as possible to achieve better energy efficiency in residential buildings to contribute to resilience and sustainability. Findings have also revealed a lack of objective and subjective data on occupants' behavior towards energy efficiency in residential buildings. In response to these gaps, the current paper has proposed a conceptual framework for occupant behavior toward a modification of thermal comfort to reduce energy use. Based on the findings of this paper, understanding the variety of factors influencing occupants' behavior should be considered a major influential factor in the design and retrofit of residential buildings with a view toward long-term resilience and sustainability.

Keywords: energy performance gap; occupant behavior; residential buildings; energy efficiency; sustainability; resilience

Citation: Far, C.; Ahmed, I.; Mackee, J. Significance of Occupant Behaviour on the Energy Performance Gap in Residential Buildings. *Architecture* **2022**, *2*, 424–433. https://doi.org/10.3390/architecture2020023

Academic Editor: Miguel Amado

Received: 11 May 2022
Accepted: 30 May 2022
Published: 2 June 2022

Publisher's Note: MDPI stays neutral with regard to jurisdictional claims in published maps and institutional affiliations.

1. Introduction

Occupant behavior is nowadays acknowledged as a main source of discrepancy between predicted and actual building performance; therefore, researchers attempt to model occupants' presence and adaptive actions more realistically [1,2]. A study by Pereira and Ramos (2019) [1] concluded that a key aspect in developing more advanced energy-efficient buildings is to consider occupant behavior in the whole development process. Many studies have demonstrated the substantial potential of efficient energy saving in buildings by considering occupant behavior in order to contribute to sustainability and resilience [2]. According to Carlucci et al. (2020) [3], one of the main sources of uncertainty in determining the actual energy consumption of buildings is occupant behavior. There is a lack of research that adequately addresses the different aspects of occupant behavior as well as the latest developments in this field [3,4]. This issue will be explored in this paper. Gunay et al. (2013, p. 234) [5] highlighted that "occupant behavior is one of the major factors influencing the energy efficiency of buildings and contributing to uncertainty in building energy use prediction and simulation". In Australia, cooling and heating contribute to almost 40% of the building's total energy consumption [6], and achieving energy conservation goals by focusing purely on technological advancements without considering building occupants' energy-related behavior will not be enough to support resilience to adapt to and mitigate climate change [7].

Several studies [8–12] have concluded that whilst almost half of the building's energy consumption is related to building envelope characteristics and utilization of the building equipment, the crucial remaining half is influenced by occupant behavior. As a result, the challenge of properly achieving energy reduction targets in buildings as a means of contributing to resilience and sustainability is partly technological and partly related to human behavior [10,11]. Several other studies [13,14] highlighted the importance of increasing people's environmental awareness to improve their thermal comfort leading to a reduction in energy consumption. Gupta and Chandiwala (2010, p. 126) [15] defined the energy performance gap as " . . . the gap between expected and actual thermal energy consumption in buildings that highly depends on the human factor", which points to the energy use implications of occupant behavior. At the moment, the understanding of occupant behavior in building design is inadequate, which leads to unrealistic assumptions and simplifications in building modeling and analysis [16,17]. The lack of reliable building occupant data has been suggested by many studies [16–20] as a significant reason for the limited understanding of the impact of occupant behavior on energy conservation and mitigation of greenhouse gas emissions [20]. As a result, there is a strong need to obtain occupant behavior data in order to develop strategies to reduce the energy performance gap as much as possible to achieve long-term climate resilience and sustainability.

As the first step in addressing this need, this paper has focused on the importance of considering the influence of occupant behavior on the energy performance gap in residential buildings to achieve the highest level of thermal performance in buildings. Residential buildings account for 70% of building floor area around the globe, while the condition and efficiency of a large part of the residential stock still need attention [20,21]. In order to close the energy performance gap as much as possible in order to achieve better energy efficiency in residential buildings, this paper has proposed a conceptual framework for occupant behavior towards the modification of thermal comfort. This paper has also recognized the absence of comprehensive occupant data as the major reason for finding limited evidence by other researchers for the impact of occupant behavior on the energy performance gap with a view toward long-term resilience and sustainability.

This paper has adopted a narrative literature review methodology to answer the main research question: What is the significance of occupant behavior in reducing the energy performance gap of residential buildings? In this approach, information is collected and interpreted with reflective summaries of findings, with the literature being described and analyzed from a contextual and theoretical point of view. To answer the research question, first, the importance and effects of occupant behavior on the amount of energy that can be saved in buildings have been presented in Section 2. Section 3 proposes a conceptual framework for occupant behavior towards the modification of thermal comfort to reduce energy use to close the energy performance gap as much as possible to achieve better energy efficiency in residential buildings, and the final conclusions are drawn in Section 3 of this paper.

Energy consumption in buildings is influenced by both building-related factors and occupant-related factors, and the latter can apply as much as or have even more impact than the former [21]. In other words, human dimensions act as significantly as technological measures in maximizing the energy performance of buildings [22]. Several studies [8–11,16] have concluded that occupant behavior has a direct and significant impact on building energy consumption. Further studies [23–25] have shown that considering occupant behavior in the building development process is as important as the quality of the building envelope and equipment.

A Cambridge University (UK) study proposed that attention to occupant behavior change can increase the energy savings in buildings by 200% [26]. Educating occupants of residential buildings could increase satisfaction with internal conditions [27]. Pilkington et al. (2011) [28] claimed that every single occupant could have a significant effect on energy efficiency and environmental impact.

2. Significance of Building Occupant Behavior on Thermal Energy Consumption in Residential Buildings and Reducing the Energy Performance Gap

A vast majority of people believe that corporations and the general public have to act much further to safeguard the environment for the sake of future generations [29]. The majority of people in different societies are not sure whom to rely on to make informed decisions to protect the environment. Trust in varied societies depends upon several factors, including people's ideas about whether or not the organization in question is honest in their communications and information or if they are truly concerned about keeping the environment safe and fair when it comes to making their own decisions toward public interests [30]. According to Vasi (2009) [29], it is realistic to consider people as bounded rationally, as they do not go through every detail when it comes to making a decision.

When discussing energy saving and protecting our planet, many researchers agree on few issues. For instance, Chen et al. (2015) [30] claimed that the energy efficiency of buildings is related to different factors such as climatic location characteristics, HVAC system, and, more importantly, occupant behavior and activities towards energy utilization. It is important to comprehend the connection between the building user's lifestyle and their energy consumption behavior to maintain or increase the building's energy efficiency [30].

There are only limited studies in the field of economics that have focused on the connection between human behavior and energy consumption [31,32]. Fischer et al. (2012) [33] presented societal change priorities in terms of a pyramid (Figure 1). Based on their proposed pyramid, applying changes to the categories at the top of the pyramid (e.g., management and policy actions) can be faster and easier but less profound compared to the categories at the bottom of the pyramid (e.g., personal and community values). They pointed out that all the social sectors, regardless of their place in the pyramid, need to collaborate on those changes simultaneously to enable the momentum for social change to spread up and down the pyramid. Ultimately, intense changes are going to be essential to make human behavior sustainable.

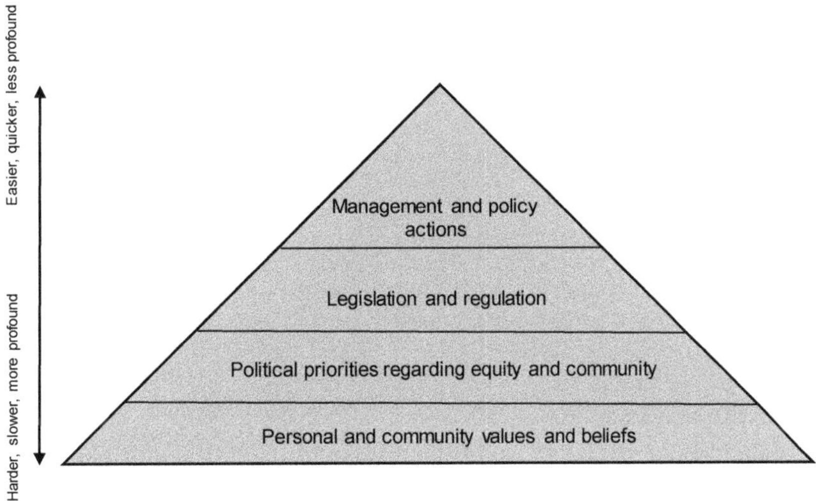

Figure 1. Societal change priorities in terms of a pyramid (after Fischer et al. 2012) [33].

According to Vasi (2009) [29], a survey showed that the public is generally concerned about environmental issues and energy consumption. Guerra-Santin and Itard (2010) [34] established a bond between energy consumption, building characteristics, and occupant behavior. As shown in Figure 2, the finding of their study revealed that analyzing household types and building characteristics in conjunction with occupant behavior result in a better understanding of the influence of the occupant on actual energy consumption and the energy-performance gap. They also found out that household characteristics, as well as the occupants' perceptions, values, and motives, are major determinants of the occupant behavior in relation to energy consumption. Energy-saving advice can also be tailored to specific household types.

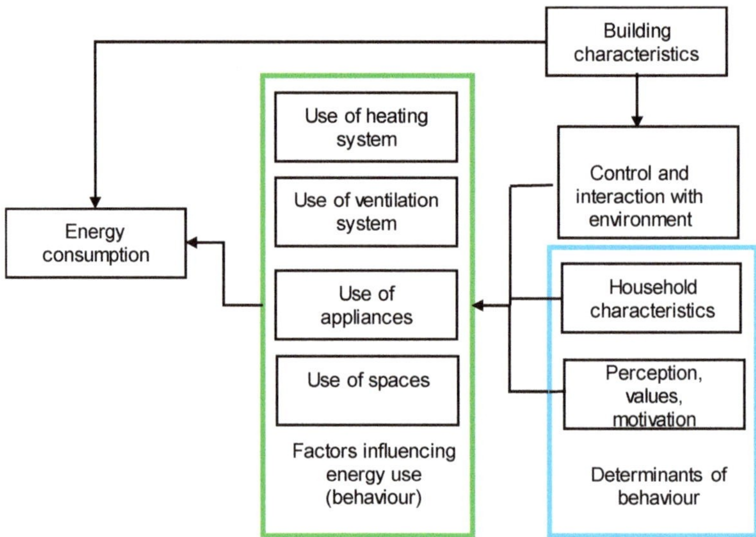

Figure 2. Relationship between energy consumption, building characteristics and occupant behavior (after Guerra-Santin and Itard 2010 [34]).

Several studies [14,35,36] highlighted the importance of increasing people's environmental awareness to improve thermal comfort for them, leading to a reduction in energy consumption.

Several studies [1,2,6,37,38] have stated that "the most highlighted gap in the literature is related to understanding the underlying social norms and priorities as well as the behaviors of households when approaching modification of thermal comfort to increase energy efficiency and sustainability in their homes." Therefore, it is important to study and understand those priorities, behaviors, and norms since they are firmly connected to particular sustainable and efficiency-related outcomes [6]. A number of researchers [38–40] specifically recommended the collection of in-depth data in the area of occupant behavior in future work. According to Carlucci et al. (2020) [3], one of the main sources of uncertainty in determining the actual energy consumption of buildings is occupant behavior. There is a lack of research that thoroughly covers the different aspects of occupant behavior as well as the latest developments in this field [4]. The need for data collection and interpretation on building occupant behavior in relation to energy consumption has been explicitly highlighted in the current literature [1,17,18].

3. Proposed Conceptual Framework

As highlighted in the previous sections, there is a strong need to obtain appropriate occupant behavior data to develop strategies for closing the energy performance gap as much as possible to achieve the highest level of thermal performance in residential buildings. In order to address this need, this section proposes a conceptual framework for occupant behavior towards the modification of thermal comfort to reduce energy use. According to Krajhanzl (2010, p. 252) [41], "a pro-environmental behavior (PEB) is defined as an individual's behavior that is generally or according to the knowledge of environmental science, judged in the context of the considered society as a protective way of environmental behavior or a tribute to the healthy environment". PEB implies taking actions that reduce the damage to the environment [42]. Kollmuss and Agyeman (2002, pp. 239–240) [43] defined pro-environmental behavior as "a type of behavior that a person deliberately chooses in order to minimize the negative impact of their actions on the environment". In the past couple of years, the common research topic for many researchers has been the motives and barriers to adopting PEB by people [41,42].

3.1. Theory of Planned Behavior (TPB)

Ajzen (1991, p. 118) [44] indicated that "the Theory of Planned Behavior is a widely accepted behavioral model for explaining and predicting behaviors through considering three core constructs: attitudes, subjective norms and perceived behavioral control". These three constructs have been the front and center of the developed conceptual framework in this study. Armitage and Conner (2001) [45] mentioned that according to 'Web of Science' core collection, TPB is one of the most focused and widely used studied behavioral models. It has been the theoretical basis for 1311 research works since 1985 [46]. Considering the merits of the Theory of Planned Behavior (TPB), this theory has been explored as the conceptual framework's theoretical underpinning in this study.

3.2. Developed Theoretical Frameworks Based on TPB for Studying Occupant Behavior

Adopting TPB as the theoretical basis, Blok et al. (2015) [47] carried out a comprehensive study on employees in the workplace at Wageningen University, the Netherlands. By adopting an online survey, people were invited by email to take participate. A total of 411 participants with equal gender distribution took part in the survey. After collecting and analyzing the data using Stata v12.1 software, Blok et al. (2015) [47] pointed out that TPB is fully capable of explaining PEB in the workplace. They have also concluded that the most significant factor in determining PEB in the workplace is the intention to act. Similarly, Belafi and Reith (2018) [48] studied occupant behavior with respect to energy use in six universities across Hungary, adopting the Theory of Planned Behavior (TPB). The result of the research highlighted the lack of motivating factors to act confidently toward pro-environmental behavior in the workplace.

Leew et al. (2015) [49] carried out a similar study based on TPB on high school students. Using survey questionnaires, they targeted high schools in Luxembourg. Analyzing the survey data, they highlighted that the attitudes toward PEB and intention to act had been the most significant factors affecting the pro-environmental behavior of the students.

3.3. A Conceptual Framework

All the mentioned studies in Section 3.2 have adopted TPB to understand occupant behavior in the education sector in Europe and carried out functional research in the area of PEB. Therefore, to study occupant behavior in residential buildings, this paper has adapted the methodology and framework used by Blok et al. (2015) [47] and Leeuw et al. (2015) to develop a conceptual framework.

In this paper, according to TPB developed by Ajzen (1991) [44], three core constructs of TPB, including attitudes, subjective norms (SNs), and perceived behavioral control (PBC), are utilized to examine building occupant behavioral intention to act pro-environmentally. Several researchers (e.g., Ajzen 1991 [44]; Blok et al. 2015 [47]; Leeuw et al. 2015 [49]) have pointed out that these three major constructs directly influence the behavioral intention to act pro-environmentally.

Figure 3 illustrates the conceptual framework diagram for occupant pro-environmental behavior towards modifying thermal comfort to reduce energy use.

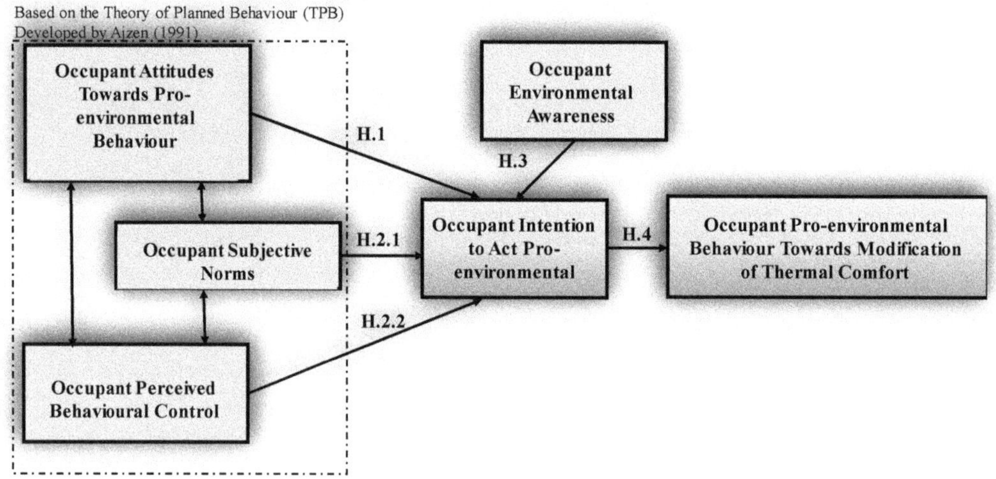

Figure 3. Proposed conceptual framework.

In the presented conceptual framework, the relationship between different items has been established using the following hypotheses:

Hypothesis 1. *Attitudes towards behaving pro-environmentally has positive impact on the occupant intention to act pro-environmental.*

Several researchers [50–53] have demonstrated that attitudes towards a certain behavior can have a constructive and direct impact on the behavioral intention of individuals. The same is applicable to the context of building occupant behavior.

Hypothesis 2.1. *Subjective norms positively influence the occupant intention to act pro-environmental.*

Subjective norms are defined in the literature as a community force to involve or not to involve in some behaviors [51]. In practical terms, the endorsement or condemnation of surrounding people creates a strong motivation for the intention to act. Synodinos and Bevan-Dye (2014) [53] stated that subjective norms have a strong and positive impact on encouraging energy conservation. For instance, the family members can encourage individuals' energy-saving behavior, such as checking whether the thermostats are set correctly at home every time they feel cold or hot.

Hypothesis 2.2. *The occupant intention to act pro-environmental is positively impacted by occupant perceived behavioral control.*

According to Ajzen (1991) [44], people's behavior is regularly subjected to the existence of means (e.g., time, money, knowledge). Those means function as constraints for the adoption of certain behaviors. Schultz and Oskamp (1996) [54] mentioned that people with high perceived behavioral control are more inclined to behave pro-environmentally.

Hypothesis 3. *Awareness of negative consequences of not behaving in pro-environmental manner encourages the occupant intention to act pro-environmental.*

Understanding the negative consequences of not behaving pro-environmentally is defined as environmental awareness [55]. Kollmuss and Agyeman (2002) [43] and Harland et al. (2007) [56] showed that awareness of the necessity of adopting pro-environmental behavior (e.g., conservation of energy), motivates individuals to be more receptive to behave pro-environmentally.

Hypothesis 4. *Intention to act pro-environmental has positive impact on PEB towards modification of thermal comfort to reduce energy use.*

The relationship between the intention to act pro-environmentally and its impacts on PEB has been studied by several researchers [53,55,56]. Those studies have indicated that the intention to act pro-environmentally has a positive impact on PEB towards energy conservation. As a result, the intention to act pro-environmentally can also have constructive effects on PEB towards the modification of thermal comfort to reduce energy use.

4. Conclusions

Based on several studies, it is understood that one of the main sources of uncertainty in determining the actual energy consumption of buildings is occupant behavior. By adopting a narrative literature review methodology, this paper highlighted the significance of considering occupant behavior to close the energy performance gap to contribute to climate resilience and sustainability. There is a lack of research that thoroughly covers the different aspects of occupant behavior as well as the latest developments in this field. Several studies indicated that whilst almost half of the residential building's energy consumption is related to building envelope characteristics and utilized equipment, the crucial remaining half is influenced by occupant behavior. As a result, the challenge of properly achieving energy reduction targets in residential buildings is partly technological and partly human behavior related.

This paper has also recognized the absence of comprehensive occupant data as the major reason for finding limited evidence by other researchers for the impact of occupant behavior on the energy performance gap. As a result, there is a strong need to obtain appropriate occupant behavior data to develop strategies for closing the energy performance gap as much as possible to achieve the highest level of thermal performance in buildings with a view toward long-term climate resilience and sustainability.

In order to address this need, this paper has proposed a conceptual framework for occupant behavior towards the modification of thermal comfort to reduce energy use. The authors of this paper are gathering and studying occupant behavior data related to energy use in residential buildings to understand the nature of occupant behavior leading toward an intention to modify thermal comfort to reduce energy consumption related to residential buildings and to develop a framework that highlights the elements of occupant behavior that can be addressed to encourage pro-environmental behavior. Employing and interpreting the occupant behavior data related to energy use in residential buildings, the proposed conceptual framework in this study will be numerically validated and further developed. The validated and developed framework and the obtained data will be published in the subsequent papers by the authors.

It is highly recommended that understanding the variety of factors influencing occupants' behavior should be considered a major influential factor in the design and retrofit of residential buildings. By obtaining appropriate occupant behavior data and proposing appropriate changes in the daily life routines and behavior of the occupants, the energy performance gap can be significantly reduced. Consequently, long-term support for climate resilience and sustainability will be provided.

Author Contributions: Conceptualization, C.F., I.A. and J.M.; methodology, C.F., I.A. and J.M.; writing—original draft preparation, C.F.; writing—review and editing, I.A. and J.M.; supervision, I.A. and J.M. All authors have read and agreed to the published version of the manuscript.

Funding: This research received no external funding.

Institutional Review Board Statement: Not applicable.

Informed Consent Statement: Not applicable.

Data Availability Statement: Not applicable.

Conflicts of Interest: The authors declare no conflict of interest.

References

1. Pereira, P.; Ramos, N. Occupant behavior motivations in the residential context–An investigation of variation patterns and seasonality effect. *Build. Environ.* **2019**, *148*, 360–1323. [CrossRef]
2. Yu, C.; Du, J.; Pan, W. Improving accuracy in building energy simulation via evaluating occupant behaviors: A case study in Hong Kong. *Energy Build.* **2019**, *202*, 378–7788. [CrossRef]
3. Carlucci, S.; De Simone, M.; Firth, S.; Kjærgaard, M.; Markovic, R.; Rahaman, M.; Annaqeeb, K.; Biandrate, S.; Das, A.; Dziedzic, J.; et al. Modelling occupant behavior in buildings. *Build. Environ.* **2020**, *174*, 360–1323. [CrossRef]
4. Mansoury, B.; Tabatabaiefar, H.R. Application of Sustainable Design Principles to Increase Energy Efficiency of Existing Buildings. *Build. Res. J.* **2016**, *6*, 167–178. [CrossRef]
5. Gunay, H.B.; O'Brien, W.; Beausoleil-Morrison, I. A critical review of observation studies, modelling, and simulation of adaptive occupant behaviors in offices. *Build Environ.* **2013**, *70*, 31–47. [CrossRef]
6. Hu, S.; Yan, D.; Azar, E.; Guo, F. A systematic review of occupant behavior in building energy policy. *Build. Environ.* **2020**, *175*, 106807. [CrossRef]
7. Zhang, Y.; Bai, X.; Mills, F. Characterizing energy-related occupant behavior in residential buildings: Evidence from a survey in Beijing, China. *Energy Build.* **2020**, *2014*, 0378–7788. [CrossRef]
8. Fabi, V.; Andersen, R.V.; Corgnati, S.; Olesen, B.W. Occupants' window opening behavior: A literature review of factors influencing occupant behavior and models. *Build Environ.* **2012**, *58*, 188–198. [CrossRef]
9. Andersen, R.K. The influence of occupant's behavior on energy consumption investigated in 290 identical dwellings and in 35 apartments. In Proceedings of the 10th International Conference on Healthy Buildings-Brisbane Convention & Exhibition Centre, Brisbane, Australia, 8–12 July 2012.
10. Lopes, M.A.R.; Antunes, C.H.; Martins, N. Energy behaviors as promoters of energy efficiency: A 21st century review. *Renew. Sustain. Energy* **2012**, *16*, 95–104. [CrossRef]
11. Ingle, A.; Moezzi, M.; Lutzenhiser, L.; Diamond, R. Better home energy audit modelling: Incorporating inhabitant behaviors. *Build. Res. Inf.* **2014**, *42*, 409–421. [CrossRef]
12. Far, C.; Far, H. Improving Energy Efficiency of Existing Residential Buildings Using Effective Thermal Retrofit of Building Envelope. *Indoor Built Environ.* **2019**, *28*, 744–760. [CrossRef]
13. Du, J.; Pan, W.; Yu, C. In-situ monitoring of occupant behavior in residential buildings–a timely review. *Energy Build.* **2020**, *212*, 0378–7788. [CrossRef]
14. Suna, C.; Zhanga, R.; Sharplesc, S.; Hana, Y.; Zhanga, H. A longitudinal study of summertime occupant behavior and thermal comfort in office buildings in northern China. *Build. Environ.* **2018**, *143*, 404–420. [CrossRef]
15. Gupta, R.; Chandiwala, S. Understanding occupants: Feedback techniques for large-scale low-carbon domestic refurbishments. *Build. Res. Inf.* **2010**, *38*, 530–548. [CrossRef]
16. Hong, T.; Taylor-Lange, S.C.; D'Oca, S.; Yan, D.; Corgnati, S.P. Advances in research and applications of energy-related occupant behavior in buildings. *Energy Build* **2016**, *116*, 694–709. [CrossRef]
17. Bordass, B.; Leaman, A.; Ruyssevelt, P. Assessing building performance in use of Conclusions and implications. *Build. Res. Inf.* **2001**, *29*, 144–157. [CrossRef]
18. Visscher, H.; Meijer, F.; Majcen, D.; Itard, L. Improved governance for energy efficiency in housing. *Build. Res. Inf.* **2016**, *44*, 552–561. [CrossRef]
19. Heydarian, A.; McIlvennie, C.; Arpan, L.; Yousefi, S.; Syndicus, M.; Schweiker, M.; Jazizadeh, F.; Rissetto, R.; Pisello, A.; Piselli, C.; et al. what drives our behaviors in buildings? A review on occupant interactions with building systems from the lens of behavioral theories. *Build. Environ.* **2020**, *179*, 360–1323. [CrossRef]
20. Far, H.; Far, C. Long-Term Structural Behavior of Composite Sandwich Panels. In Proceedings of the 2nd World Congress on Civil, Structural, and Environmental Engineering (CSEE'17), Barcelona, Spain, 2–4 April 2017.
21. Yoshinoa, H.; Hong, T.; Nord, N. IEA EBC annex 53: Total energy use in buildings—Analysis and evaluation methods. *Energy Build.* **2017**, *152*, 124–136. [CrossRef]
22. D'Oca, S.; Hong, T.; Langevin, J. The human dimensions of energy use in buildings: A review. *Renew. Sustain. Energy Rev.* **2018**, *81*, 731–742. [CrossRef]

23. Gram-Hanssen, K. Efficient technologies or user behavior, which is the more important when reducing households' energy consumption. *Energy Effic.* **2013**, *6*, 47–57. [CrossRef]
24. Schweiker, M.; Shukuya, M. Comparative effects of building envelope improvements and occupant behavioral changes on the exergy consumption for heating and cooling. *Energy Policy* **2010**, *38*, 76–86. [CrossRef]
25. James, M.; Ambrose, M. Retrofit or behavior change? Which has the greater impact on energy consumption in low income households. *Procedia Eng.* **2017**, *180*, 1558–1567. [CrossRef]
26. Markusson, N.; Ishii, A.; Stephens, J. The social and political complexities of learning in carbon capture and storage demonstration projects. *Glob. Environ. Chang.* **2011**, *21*, 293–302. [CrossRef]
27. Menadue, V.; Soebarto, V.; Williamson, T. Perceived and actual thermal conditions: Case studies of green-rated and conventional office buildings in the City of Adelaide. *Archit. Sci. Rev.* **2014**, *57*, 1758–9622. [CrossRef]
28. Pilkington, B.; Roach, R.; Perikins, J. Relative Benefit of Technology and Occupant Behavior in Moving Towards a More Energy Efficient Sustainable Housing Paradigm. *Energy Policy* **2011**, *39*, 4962–4970. [CrossRef]
29. Vasi, B. Social Movements and Industry Development: The Environmental Movement's Impact on the Wind Energy Industry Mobilization. *Int. J.* **2009**, *14*, 315–336. [CrossRef]
30. Chen, S.; Yang, W.; Yoshino, H.; Levine, M.D.; Newhouse, K.; Hinge, A. Definition of occupant behavior in residential buildings and its application to behavior analysis in case studies. *Energy Build.* **2015**, *104*, 1–13. [CrossRef]
31. Maréchal, K. Not irrational but habitual: The importance of behavioral lock-in in energy consumption. *Ecol. Econ.* **2010**, *69*, 1104–1114. [CrossRef]
32. Galassi, V.; Madlener, R. Shall I open the window? Policy implications of thermal-comfort adjustment practices in residential buildings. *Energy Policy* **2018**, *119*, 518–527. [CrossRef]
33. Fischer, J.; Dyball, R.; Fazey, I.; Gross, C.; Dovers, S.; Ehrlich, P.; Brulle, R.; Christensen, C.; Borden, R. Human behavior and sustainability. *Front. Ecol. Environ.* **2012**, *10*, 153–160. [CrossRef]
34. Guerra-Santin, O.; Itard, L. Occupants' behavior: Determinants and effects on residential heating consumption. *Build. Res. Inf.* **2010**, *38*, 318–338. [CrossRef]
35. Eon, C.; Morrison, G.; Byrne, J. The influence of design and everyday practices on individual heating and cooling behavior in residential homes. *Energy Effic.* **2018**, *11*, 273–293. [CrossRef]
36. Frontczak, M.; Andersen, R.; Wargocki, P. Questionnaire survey on factors influencing comfort with indoor environmental quality in Danish housing. *Build. Environ.* **2012**, *50*, 56–64. [CrossRef]
37. Stevenson, F. Leaman, Evaluating housing performance in relation to human behavior: New challenges. *Build. Res. Inf.* **2010**, *38*, 437–441. [CrossRef]
38. Judson, P.; Maller, C. Housing renovations and energy efficiency: Insights from homeowners' practices. *Build. Res. Inf.* **2014**, *42*, 501–511. [CrossRef]
39. Moy, C. Rainwater Tank Households: Water Savers or Water Users? *Geogr. Res.* **2012**, *50*, 204–216. [CrossRef]
40. Willand, N.; Horne, R. Low Carbon Residential Refurbishments in Australia: Progress and prospects'. In Proceedings of the State of Australian Cities Research Network, Australia, 2013; Available online: http://apo.org.au/resource/low-carbon-residential-refurbishments-australiaprogress-and-prospects (accessed on 9 February 2021).
41. Krajhanzl, J. Environmental and pro-environmental behavior. *Sch. Health* **2010**, *21*, 251–274.
42. Steg, L.; Vlek, C. Encouraging pro-environmental behavior. *J. Environ. Psychol.* **2009**, *29*, 309–317. [CrossRef]
43. Kollmuss, A.; Agyeman, J. Mind the Gap: Why do people act environmentally and what are the barriers to pro-environmental behavior? *Environ. Educ. Res.* **2002**, *8*, 239–260. [CrossRef]
44. Ajzen, I. The theory of planned behavior. *Organ. Behav. Decis. Process.* **1991**, *50*, 179–211. [CrossRef]
45. Armitage, C.J.; Conner, M. Efficacy of the Theory of Planned Behavior: A Meta-Analytic Review. *Br. J. Soc. Psychol.* **2001**, *40*, 471–499. [CrossRef]
46. Wu, Z.; Wang, B.; Xia, X. Large-scale building energy efficiency retrofit: Concept, model and Control. *Energy* **2016**, *109*, 456–465. [CrossRef]
47. Blok, V.; Wesselink, R.; Studynka, O.; Kemp, R. Encouraging sustainability in the workplace: A survey on the pro-environmental behavior of university employees. *Renew. Sustain. Energy Rev.* **2015**, *52*, 960–975. [CrossRef]
48. Belafi, Z.; Reith, A. Interdisciplinary survey to investigate energy related occupant behavior in offices- the Hungarian case. *Int. J. Eng. Inf. Sci.* **2018**, *13*, 41–52.
49. Leeuw, A.; Valois PAjzen, I.; Schmidt, P. Using the theory of planned behavior to identify key beliefs underlying pro-environmental behavior in high-school students: Implications for educational interventions. *J. Environ. Psychol.* **2015**, *42*, 128–138. [CrossRef]
50. Fishbein, M.; Ajzen, I. *Belief, Attitude, Intention, and Behavior: An Introduction to Theory and Research*; Addison-Wesley: Reading, MA, USA, 1975.
51. Ajzen, I.; Fishbein, M. The influence of attitudes on behavior. *Handb. Attitudes* **2005**, *173*, 221.
52. Clement, C.A.; Henning, J.B.; Osbaldiston, R. Integrating Factors that Predict Energy Conservation: The Theory of Planned Behavior and Beliefs about Climate Change. *J. Sustain. Dev.* **2014**, *7*, 46–69. [CrossRef]
53. Synodinos, C.; Bevan-Dye, A. Determining African Generation Y Students' Likelihood of Engaging in Pro-environmental Purchasing Behavior. *Mediterr. J. Soc. Sci.* **2014**, *5*, 101–110.

54. Schultz, P.W.; Oskamp, S. Effort as a moderator of the attitude-behavior relationship: General environmental concern and recycling. *Soc. Psychol. Q.* **1996**, *59*, 375–383. [CrossRef]
55. Schwartz, S.H. Normative influences on altruism. *Adv. Exp. Soc. Psychol.* **1977**, *10*, 221–279.
56. Harland, P.; Staats, H.; Wilke, H.A. Situational and personality factors as direct or personal norm mediated predictors of pro-environmental behavior: Questions derived from norm-activation theory. *Basic Appl. Soc. Psychol.* **2007**, *29*, 323–334. [CrossRef]

Article

Exploring the Early Impacts of the COVID-19 Pandemic on the Construction Industry in New York State

Esther Ilatova [1,*], Yewande S. Abraham [1,*] and Bilge Gokhan Celik [2]

1   Department of Civil Engineering Technology Environmental Management and Safety,
    Rochester Institute of Technology, Rochester, NY 14623, USA
2   School of Engineering, Computing and Construction Management, Roger Williams University,
    Bristol, RI 02906, USA; bcelik@rwu.edu
*   Correspondence: ei6742@g.rit.edu (E.I.); ysaite@rit.edu (Y.S.A.)

Abstract: The COVID-19 pandemic severely impacted many industries on a global scale. Expectedly, the construction industry was not left out as non-essential construction was halted, strict health and safety protocols were introduced, and businesses were disrupted. New York City was the epicenter of the pandemic at its onset in the United States, and the pandemic had different impacts on workers based on their work location and role. This study utilized a survey including twenty-five statements to explore the initial impacts of the COVID-19 pandemic on the construction industry in New York State, analyzing its effects on sixty-one construction industry professionals, their projects, and firms, also considering their work location and role in the construction process. The most severe impacts were on construction schedules and in-person meetings. Those who worked in New York City had more difficulty complying with the increased health and safety regulations than those who worked outside the city. Those categorized as builders indicated significantly more contract performance issues. Furthermore, a set of recommendations were highlighted to strengthen the industry's response to future similar disruptions. This study is significant in helping researchers and businesses build more resilient operations to address current and future pandemic-related challenges facing the construction industry.

Keywords: COVID-19 pandemic; construction industry; health and safety; communication; construction projects

Citation: Ilatova, E.; Abraham, Y.S.; Celik, B.G. Exploring the Early Impacts of the COVID-19 Pandemic on the Construction Industry in New York State. *Architecture* **2022**, *2*, 457–475. https://doi.org/10.3390/architecture2030026

Academic Editors: Iftekhar Ahmed, Masa Noguchi, David O'Brien, Chris Tucker and Mittul Vahanvati

Received: 1 June 2022
Accepted: 26 June 2022
Published: 29 June 2022

1. Introduction

The ongoing pandemic has adverse effects on individuals, families, businesses, organizations, and governments globally. Studies concluded that the economic decline of 2020 caused by the COVID-19 pandemic is the deepest since the Second World War [1,2]. The pandemic's effect on the global economy is disruptive, affecting production, supply chains, and firms [3,4]. Many uncertainties are associated with economic recovery after the pandemic [1,5]. Similar to the impact of natural disasters, the pandemic led to a decrease in business operations. Impoverished neighborhoods were severely hit, and grocery store sales experienced a surge due to a higher demand for necessities [6]. Additionally, manufacturers of non-essential goods experienced fewer orders, and oil prices plummeted as non-essential travel decreased [3,6]. Many of the consequences of the pandemic directly impact the construction industry, which contributes approximately 5% of the overall US GDP [7] (contributed 4.1% in 2019) and 3.1% of the GDP in New York [8]. The construction industry is one of the largest globally [9], and it continues to play a critical role in revitalizing the US economy [10].

Evidently, the COVID-19 pandemic inflicted damage resulting in higher costs, widespread closures, and lockdowns. Furthermore, it has deeply impacted the world in many aspects, such as economies, healthcare systems, and travel and leisure [11]. New York State, particularly New York City, was severely hit at the onset of the pandemic, with New York City becoming an epicenter accounting for hundreds of deaths daily between March and May 2020 [12,13]. New York City held 39.2% of construction industry jobs in New York state in 2019 and had 52.2% of construction job losses in the state [14]. In considering the rates of COVID-19 outbreaks by industry, construction accounted for the second or third highest rates of outbreaks in Washington, Michigan, Utah, and Nashville, Tennessee [15]. The impact of the pandemic is far-reaching, and it continues to affect many lives and sources of livelihood. The full extent of the impact of the ongoing pandemic is still unknown. According to the International Monetary Fund (IMF), "quarantines, regional lockdowns, and social distancing—which are essential to contain the virus—curtail mobility, disrupt supply chains and lower productivity." In addition, "layoffs, income declines, fear of contagion . . . and heightened uncertainty . . . [trigger] further business closures and job losses" [16] (p. 17). While a few studies have focused on the pandemic's impact on the construction industry, there is a scarcity of research focusing on challenges directly associated with construction business management, construction projects, and the workforce. Additionally, since the nature of the impacts of the pandemic on construction projects and workers is primarily influenced by the location of construction businesses and their assigned projects [17], there is a need to consider differences in the impacts of COVID-19 on the construction industry in different locations and also consider the roles of construction industry professionals.

The overarching research question is as follows: what is the impact of the COVID-19 pandemic on the construction industry in New York State? In addition, this study addressed two additional questions: (1) were certain groups more adversely affected based on: (a) working primarily in or out of New York City; (b) their role in the construction process as builders or non-builders? The second question was: (2) what can the construction industry do to better prepare for future disruptions?

The remaining sections of this paper are organized as follows. First, a background into the impact of COVID-19 on the construction industry considering the main themes of this study is provided, followed by the research methodology, the results, the discussion of the findings, and the conclusion and limitations.

2. Background

2.1. Impact of COVID-19 on the Construction Industry

Compared to other industries, such as hospitality, the impact of COVID-19 on the construction industry was arguably less severe in 2020 [11,18]. This is primarily because most construction was deemed an essential service [19]. In New York State, essential construction services continued during the pandemic. Several definitions qualify as essential construction under New York State Executive Order 202.6 [20]. One definition is "construction necessary to protect the health and safety of occupants of a structure" [20]. Another definition is "construction necessary to continue a project if allowing the project to remain undone would be unsafe, provided that construction must be shut down when it is safe to do so" [20]. After restrictions were lifted, there was an increase in residential construction in the US. The increase in housing-related spending is partly due to more people working from home, lower interest rates, and the increased purchase of second homes [21]. Research further predicts that the construction of single-family residential projects will increase by 7%, commercial projects will increase by 5%, and manufacturing will flatline in 2021 [22]. Despite this glimmer of hope for the construction industry and the availability of government aid, many firms had to lay off workers due to their inability to pay overhead costs, make payroll, and more [18]. There was more than a 40% drop in employment in New York State between March and April 2020 [18]. Alsharef et al. [23] explored the early impact of the COVID-19 pandemic on the construction industry in the US. They indicated significant project and material delivery delays, increased costs of materials, and lower productivity rates [23]. However, the pandemic also brought about some opportunities, including increased prospects of recruiting skilled workers, an increase in the construction of residential buildings, and more medical- and transportation-related construction projects [23]. Choi and Staley [24] identified the challenges faced by the construction industry amplified by the pandemic to include regulatory confusion, vulnerable employee risk, low COVID-19 literacy, and supply-chain shortages.

2.2. Impact of COVID-19 on Construction Projects

In addition to overall impacts on the construction industry, COVID-19 also impacted project timelines and budgets due to delays in lead times [25], limited access to resources [23], supply-chain bottlenecks [24], and international restrictions on travel [26]. The impact of the pandemic extended to suppliers and vendors who could not meet up with demands and, in turn, affected the timeline of construction projects. In a survey of 53 construction contractors in New York State, 81% of them indicated that their projects experienced delays due to longer lead times or shortages of materials, while 62% cited delivery delays [27]. Furthermore, the pandemic may cause breaches in contractual and legal obligations to clients [23]. As of August 2021, the pandemic continues to impact construction projects with a knock-on effect on claims [28]. Chivilo, Fonte, and Koger [17] indicated that the nature of the impacts of the pandemic on construction projects and workers are primarily influenced by the location of the construction businesses and their assigned projects. Some of the direct impacts include project termination or the cancellation of contracts [29]. In total, 62% of contractors in New York State had their projects canceled, postponed, or scaled back for any reason [27]. Several construction activities were shut down in some states depending on whether they were considered essential or not [23].

2.3. Impact of COVID-19 on Construction Firms and Company Management

Globally, construction firms have faced difficulties during the pandemic relating to finances, the loss of projects and clients, budget adjustments, contractual and legal issues, communication, health and safety, inadequate resources, pandemic-related delays in lead times due to restricted movements, and difficulties caused by working remotely [26]. Travel bans have also contributed to material delivery delays, equipment being left on non-operational sites, factories shutting down, and suspensions in the production and distribution of supplies [10]. Such delays jeopardized the profitability of projects, triggered specific contract clauses to take effect [17], and triggered the extension of deadlines [10]. Furthermore, the pandemic caused disruptions in the scheduling and procurement of supplies and services [24,30] and has highlighted the importance of recognizing and mitigating financial and legal risks [17]. In a survey conducted by Turner and Townsend, for the UK, half of the respondents reported an "increase in contractual disputes since the start of the pandemic" [31]. In addition, "83% of respondents reported a pause or temporary site closure because of COVID-19, and 72% reported reduced productivity on projects compared to pre-pandemic levels" [31]. Moreover, 49% of respondents reported adding "up to 10% extra for COVID-19 related costs in bid submissions" [31]. In a 2021 survey of construction contractors in the US, 63% of firms passed on additional costs due to rising material costs [32]. They also reported an increase in the volume of business compared to a year earlier.

The pandemic has led to the rise of short- and long-term trends in the construction industry. The use of digital collaboration platforms has increased due to the need to work remotely [33]. In total, 57% of US contractors indicated an increase in the rate of technology adoption in their firms [32]. In particular, they reported that their firms adopted more project management software. The present short-term trends include contractors using online tools to manage and monitor their employees' productivity and wellbeing, manage resources and cash flow, and conduct online meetings [33]. In addition, contractors are looking toward finding alternative sources and stockpiling supplies and materials in the short term, particularly those with long lead times, to increase resiliency [34]. They are also making connections with new suppliers. In the long term, companies are looking into further investments in technology to increase work productivity and manage workflow [32,33]. As a result, there is an expected increase in the automation of construction and design elements. A push toward sustainability is also expected to occur as the trend for promoting healthier lifestyles is more prominent now with the pandemic than ever before [33]. This will result in an emphasis on air quality, access to outdoor space, and an increase in the use of sustainable materials [35].

2.4. Impact of COVID-19 on Health and Safety in the Construction Industry

Although many industries and businesses could explore remote working options, most of the construction industry's operations must be on-site and cannot be done remotely [26]. Therefore, the shift to working remotely can only go so far for construction firms. It is also clear that unrestricted work in high-contact industries, such as construction, is associated with a higher level of community transmission, increased risks to at-risk workers, and considerable health disparities among members of certain racial and ethnic minority groups [36]. The pandemic also impacts the mental health of construction workers [37].

The COVID-19 pandemic triggered many public health emergency standards and regulations to be implemented across all industries [23]. These standards and regulations do not replace existing local, state, and federal laws that must be observed. Rather, they are supplemental to existing compliance requirements. Some of these requirements include maintaining six feet of distance between personnel, providing personal protective equipment (PPE) such as face coverings to personnel, limiting in-person gatherings, limiting the number of people allowed to occupy tight spaces such as elevators at any given time, and having employees work from home whenever feasible [38]. Employers were also required to implement disinfectant protocols and mandatory screening assessments [38]. All of the requirements as mentioned above merely scratch the surface of what construction firms must observe. Integrating COVID-19 protocols with existing safety protocols in organizations can increase the effectiveness of the measures adopted by organizations on their projects [37]. Recommendations to manage COVID-19 in the construction industry include preventive measures to protect the health and safety of construction workers and COVID-19 education, where employers keep their workers up to date on relevant information to keep them safe [24].

2.5. Impact of COVID-19 on Communication in the Construction Industry

State and federal guidelines mandated that employees work remotely from home whenever possible [38]. They also required that in-person meetings are limited in the number of people and duration [38]. Therefore, most meetings and conferences shifted from in-person to various telecommunication platforms [39,40]. This could be burdensome for construction firms in some respects, because most of their workforce needs to be on-site. As a result, although remote work is the safer practice to preserve the health and safety of the employees, the project itself may suffer [41].

The number of employees working remotely across the country skyrocketed during the pandemic, and about 50% of US workers that were surveyed reported that they were working from home [42]. While there are many benefits associated with working remotely, such as decreasing overhead costs, reducing travel times to and from work, reducing the carbon footprint and so forth, several disadvantages must be noted [43]. For example, the online workplace can lead to increased mental health issues due to the inability to "unplug" [41]. In addition, some people must juggle the roles of being an employee and parent or caretaker working from home [43]. Furthermore, many employees "do not possess sophisticated systems along with headsets, video cameras, high-speed connectivity, and the skills to manage all" ([44], p. 10). Therefore, working from home can be overwhelming for some individuals. Employees across industries reported that the number of meetings they had on a regular basis increased after the pandemic [43]. Many claimed they experienced mental and emotional exhaustion while working remotely [23]. On the other hand, "almost 60% of Americans think COVID-19 has changed the way we work for the better" [43]. "Close to 99% of respondents ... reported that their employers are showing empathy toward staff" and "85% ... feel that their employers are doing a good job of communicating and informing staff about the company's situation and ongoing response to the pandemic" [43].

3. Methods

3.1. Research Approach

This study involved multiple steps to address the central research questions (Figure 1). First, a detailed literature search was conducted using the keywords COVID-19, construction, construction industry, construction firm, and construction management in different combinations on the Google Scholar database and Google search engine. Then, relevant articles and gray literature were identified, and the contents were categorized into seven themes: construction firm, construction project, financial/economic, project schedule, health and safety, communication, and supply chain/procurement. These themes were then narrowed to four broader impact categories: company management, construction projects, health and safety, and communication.

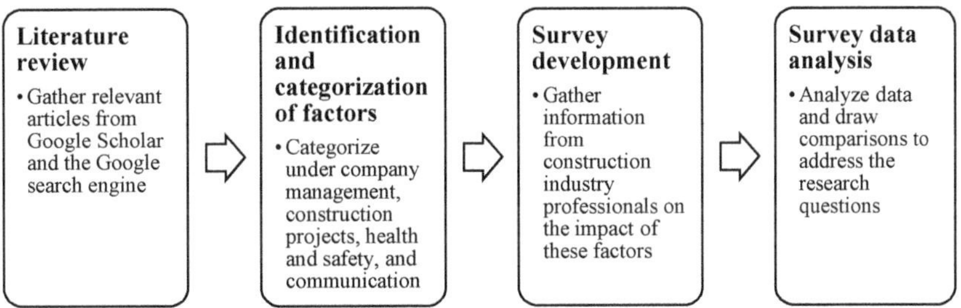

Figure 1. Research framework.

Table 1 shows sixteen of the prominent factors identified in the literature through the literature review and distributed among the four impact categories. The factors identified under company management are contracts and legal issues, lack of new contracts, company recovery time, workforce reduction, financial constraints, and new opportunities such as technology adoption. For the construction projects category, the factors were project team size, project suspension or shutdown, schedule interruption and delays, supply chain issues (i.e., material shortage and procurement delays), productivity, and project planning. Under health and safety, two factors were identified, namely COVID-19 health and safety regulation compliance and workplace safety. Finally, the factors under communication focused on virtual meetings and remote work. The identified factors were then converted into twenty-five statements included in the survey to assess the impact of the pandemic on the construction industry in New York State. Surveys were selected to gauge the views of a number of respondents within the study's geographical boundaries in order to address the research questions.

Table 1. Categorization of factors that impacted the construction industry as a result of the COVID-19 pandemic.

Category	Factors	Sources
Company Management	Contracts and legal issues	[17,23,26,31,34,41]
	Lack of new contracts	[23,29]
	Company recovery time	[29]
	Workforce reduction	[18,23,29,34,41]
	Financial constraints	[17,23,25,26,29,41]
	Technology adoption	[26,32,33]

Table 1. *Cont.*

Category	Factors	Sources
Construction Project	Project team size	[23,29]
	Project suspension or shutdown	[23,26,27,29]
	Schedule interruption and delays	[10,23,25,26,31]
	Supply chain issues	[10,23–25,27,34]
	Productivity	[23,29,31–33]
	Project planning	[23,26,41]
Health and Safety	COVID-19 health and safety regulations	[23,24,41,45]
	Workplace safety	[23,24,37,46]
Communication	Virtual meetings	[23,38,40,43]
	Remote work	[23,26,33,38,41,42]

3.2. Survey Design

This study involved a cross-sectional survey created to examine the impact of COVID-19 on the construction industry in New York State. A cross-sectional survey design provides data to draw inferences about a population of interest [47]. The study was submitted to the institutional review board at the authors' institution for approval, and the study was exempt. The participants were selected for the study based on the specified inclusion and exclusion criteria, such as actively working in the construction industry, working in New York State, and having at least one year of construction experience [46]. This study included a survey questionnaire consisting of 6 demographic questions and 25 Likert-scale statements, which measured the impact of various pandemic-related factors on the construction industry in New York State. The 25 Likert-scale statements were grouped under the four impact categories (company management, construction projects, communication, and health and safety). The participants were also asked if they could work from home, if their productivity had changed, and if their firm took advantage of government aid to mitigate the effects of the pandemic on their businesses. The factors in the Likert-scale statements were determined from the literature review, and the survey questions were pilot tested with professionals in the construction industry. Survey participants also had the opportunity to provide recommendations and share any other relevant information.

Based on Nemoto and Beglar [48], analysis has shown that Likert Scales with more than six categories are rarely tenable, possibly because of limitations on human working memory capacity. Losby and Wetmore [49] mentioned that when comparing between a four-point and five-point Likert scale, the overall difference in the responses is negligible, and the authors' decision to use an odd number scale was made to allow the respondents to select a midpoint if they were neutral about the impact of the factor. Therefore, a five-point Likert scale was selected, the questions asked participants to respond according to whether they strongly agreed (5), somewhat agreed (4), neither agreed nor disagreed (3), somewhat disagreed (2), and strongly disagreed (1) to certain pandemic-related statements. The survey data were collected through Qualtrics, an online survey development platform.

3.3. Participants

Participants were selected through the convenience and purposive sampling approaches, which are non-random sampling methods [50]. The survey was sent out to members of a selection of professional organizations in New York, distributed to construction industry professionals on LinkedIn, and through the professional networks of the researchers. The survey was limited to professionals working in the construction industry in New York State. Participants who did not work in New York State and those with no connection to the construction industry were excluded. A total of seventy-nine survey responses were received. Three respondents who answered that they do not conduct projects in New York State were excluded from this study. In addition, fifteen respondents were excluded since their surveys were less than 30% complete and were missing answers to the background qualifying questions. As a result, a total of sixty-one valid responses were collected.

The participants were asked to indicate their organization type. The majority were working as contractors or working in architecture/engineering firms (Figure 2). The participants were further categorized into whether they were builders or non-builders based on if their work was field-based or office-based. Those who identified their organization as a contractor or vendor/manufacturer were categorized as builders, while the non-builder category included those who worked for architecture and engineering firms and construction project clients. Consequently, thirty were identified as builders, while thirty-one were identified as non-builders.

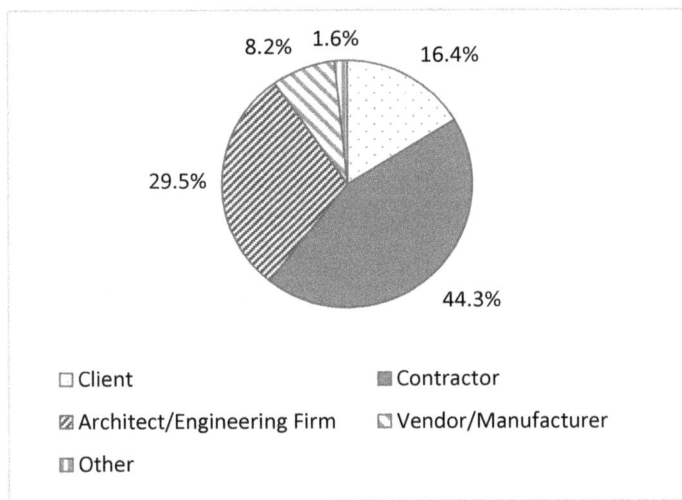

Figure 2. Participants' organization type.

The participants were also asked to indicate their role within the construction industry; 27.9% were upper management, 24.6% were engineers, 18% were project managers, 8.2% were contractors, 6.6% were architects, and 9.8% held other roles in construction, such as estimators, superintendents, project coordinators, or consultants.

The participants also indicated whether they worked for a private or government-owned organization. In total, 88.5% worked for a private organization, 4.9% worked for a government-owned organization, and 6.6% selected the "other" category. The majority of the respondents had six to ten years of experience. Approximately 85% of the participants had more than five years of construction experience (Figure 3).

The participants were asked to identify their primary work location. Since New York City was severely impacted at the initial onset of the pandemic, the study aimed to identify whether the impacts of the pandemic differed for those primarily working in New York City. The survey questions were designed to determine where the subjects conducted their primary job duties regardless of where their main offices may be located. In total, 32.8% indicated that they primarily worked in New York City, while 67.2% indicated that they worked in other parts of New York State.

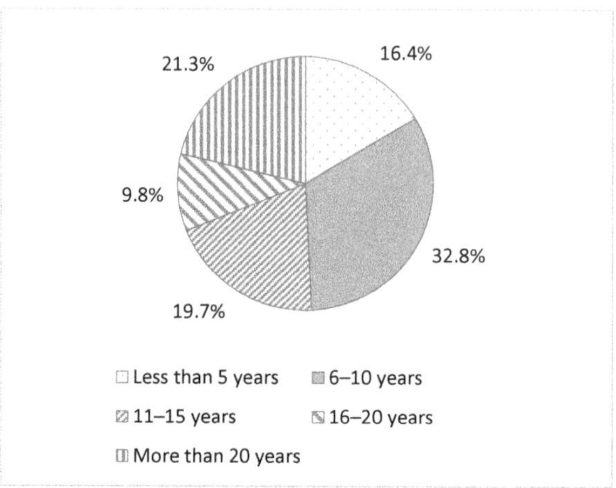

Figure 3. Years worked in the construction industry.

3.4. Data Analysis

Survey data were downloaded and exported from Qualtrics into Microsoft Excel. Each survey response was screened and excluded from analysis under certain circumstances, such as having incomplete responses or participants not meeting location requirements (working in New York State), or not working in construction. The data were preprocessed in Microsoft Excel and analyzed using SPSS version 26 statistical software. For analysis, descriptive statistics and independent samples t-tests were conducted. The central limit theorem holds when a sample size is greater than thirty and a study sample is considered normally distributed [51]. In addition, Mann–Whitney non-parametric tests were used when any variables did not meet the normality assumptions.

3.5. Validity and Reliability

Content validity "can be defined as the ability of the selected items to reflect the variables of the construct in the measure" [52] (p. 166). If an instrument lacks content validity, it is impossible to establish its reliability [52] (p. 166). Therefore, in addition to identifying key factors relating to the impact of COVID-19 on the construction industry in the literature, the survey questions were checked with experts in the construction industry to determine if they covered key aspects of the constructs being measured. Reliability is a concept that measures the extent to which an instrument produces consistent results [53]. Reliability was tested by determining the Cronbach's Alpha of items within the four impact categories.

3.6. Reliability Analysis for Likert-Scale Questions

Table 2 shows the four impact categories for the COVID-related impact statements and their Cronbach's Alpha scores. The impact categories include company management, construction projects, health and safety, and communication. The Cronbach's Alpha for the statements within each category was computed to determine the internal consistency of the variables. Cronbach's Alphas for the nine company management, ten construction projects, three health and safety, and two communication statements were 0.838, 0.810, 0.820, and 0.847, respectively. One of the health and safety statements was excluded to improve the reliability of that category.

Table 2. Reliability analysis for survey questions.

ID	Category	Related Impact Statements	Cronbach's Alpha
CM	Company Management	CM1: My company experienced contract performances issues due to the pandemic CM2: My company experienced legal consequences due to contractual breaches caused by the pandemic CM3: My company experienced difficulty in obtaining new projects due to the pandemic CM4: My company experienced difficulty in obtaining new clients due to the pandemic CM5: My company's recovery timeline from the pandemic is unknown CM6: The pandemic has forced my company to lay off workers CM7: My company had to cut down expenses due to the pandemic CM8: My company embraced new technology as a result of the pandemic CM9: Overall, my company benefited from changes brought upon by the pandemic	0.838
CP	Construction projects	CP10: The pandemic reduced the size of my project team CP11: The pandemic has caused active projects to be suspended CP12: The pandemic led to the shutdown of my company's project sites CP13: My projects experienced schedule interruptions due to the pandemic CP14: The pandemic caused project scheduling delays CP15: My project suffered from material shortage due to the pandemic CP16: Material procurement was delayed as a result of the pandemic CP17: My company requested extended project deadlines due to the pandemic CP18: My workload has been reduced since the start of the pandemic CP19: The pandemic increased the number of hours dedicated to project planning	0.810
HS	Health and safety	HS20: COVID-19 Health and Safety regulations were difficult to comply with HS21: The new PPE requirements were difficult to comply with HS22: I felt safer at my workplace due to the new regulations * HS23: There were high COVID-19 positivity rates on my projects	0.820
CO	Communication	CO24: In-person meetings were limited CO25: Some meetings switched to virtual platforms (i.e., Zoom, Skype, Microsoft Teams, etc.)	0.847

* Statement excluded from the HS Category due to a low Cronbach's Alpha score.

4. Results

4.1. Analysis of Overall Survey Results

The survey participants' level of agreement with certain COVID-related impact statements is presented in Figure 4. In the company management category, while 78.6% of the participants strongly agreed or somewhat agreed that their companies experienced contract performance issues, only 21.3% strongly agreed or somewhat agreed with the statement that their companies experienced legal consequences due to contractual breaches caused by the pandemic. A total of 71.2% of the participants' companies experienced difficulty obtaining new projects due to the pandemic, and 62.7% had difficulty obtaining new clients. In addition, 75.4% strongly agreed or somewhat agreed with the statement that their company had to cut down expenses, and 81.9% of the participants strongly agreed or somewhat agreed that their companies embraced new technology due to the pandemic. Finally, 54.1% somewhat agreed or strongly agreed that their company benefitted from changes brought about by the pandemic.

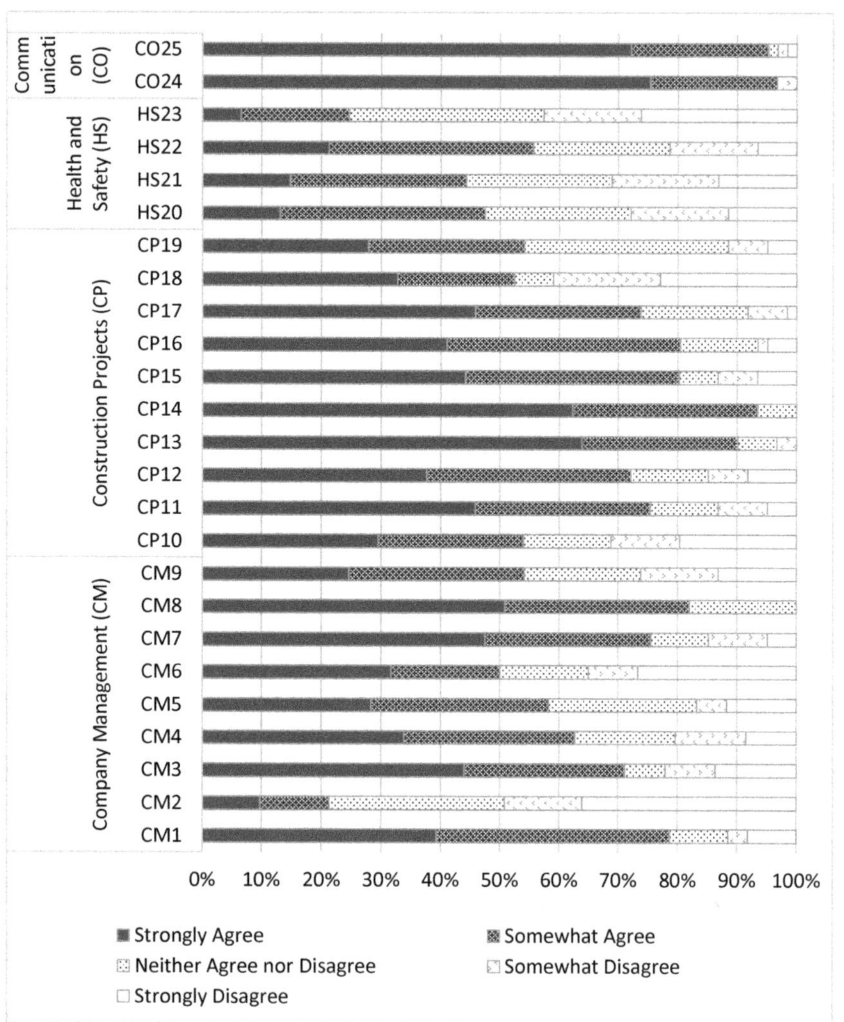

Figure 4. Responses to statements on the impact of COVID-19 on the construction industry in New York State.

In the construction projects category, 54.1% indicated that the pandemic reduced the size of their project teams, 75.4% strongly agreed or somewhat agreed that the pandemic caused active projects to be suspended, and 72.1% mentioned that the pandemic led to the shutdown of their company's project sites. A total of 93.4% strongly agreed or somewhat agreed that the pandemic caused project scheduling delays. Additionally, 80.4% experienced material shortages and procurement delays due to the pandemic. In total,

73.8% stated that their companies had to request project schedule extensions to meet their project deadlines. Over half of the respondents indicated having reduced workloads and a similar percentage stated dedicating more hours to project planning during the pandemic.

In the health and safety category, 47.5% strongly agreed or somewhat agreed with the statement that COVID-19 health and safety regulations were difficult to comply with, 44.3% indicated that the new PPE requirements were difficult to comply with, 55.7% felt safer at their workplace due to the new regulations, and 24.6% strongly agreed or somewhat agreed that there were high COVID-19 positivity rates on their projects.

In the communication category, 96.7% strongly agreed or somewhat agreed that in-person meetings were limited, and 95.1% strongly agreed or somewhat agreed that some meetings switched to virtual platforms.

The survey asked if the participants had the opportunity to work from home since the beginning of the pandemic. Overall, 65.6% stated to have worked from home, while 34.4% did not work from home. For the question regarding the impact of the pandemic on their productivity, 31.1% of the participants stated that they were less productive, and 32.8% reported that they were more productive (Figure 5). A cross-tabulation analysis revealed that more of those who worked from home reported having an increase in their productivity (40%) compared to those who did not work from home (19%). Furthermore, the authors identified a significant difference between those who worked from home and those who did not when it came to adopting new technology. Those who worked from home reported a higher rate of technology adoption (Mdn = 5.0) than those who did not work from home (Mdn = 1.0). A Mann–Whitney U test indicated that this difference was statistically significant, $U(N_{worked\text{-}from\text{-}home} = 40, N_{did\text{-}not\text{-}work\text{-}from\text{-}home} = 21) = 245.00$, $z = -2.91, p = 0.004$.

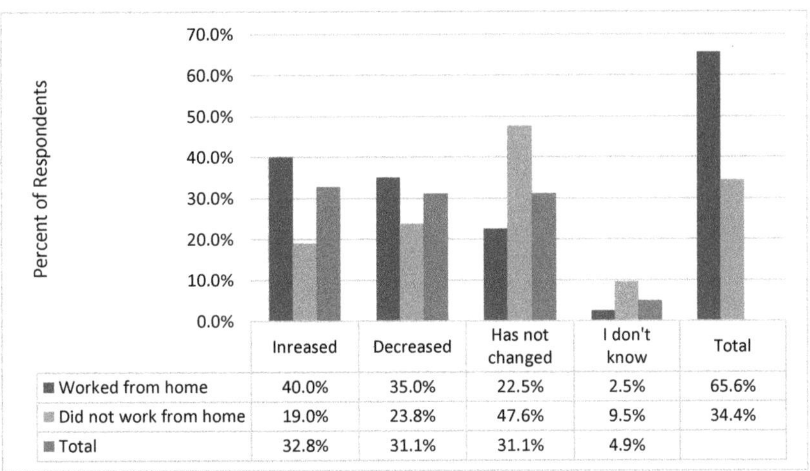

	Inreased	Decreased	Has not changed	I don't know	Total
Worked from home	40.0%	35.0%	22.5%	2.5%	65.6%
Did not work from home	19.0%	23.8%	47.6%	9.5%	34.4%
Total	32.8%	31.1%	31.1%	4.9%	

Figure 5. Impact of the pandemic on work productivity.

In response to whether their company took advantage of government aid to mitigate the pandemic's effects on their businesses, approximately half of the respondents reported that their company had taken advantage of government aid to mitigate the impact of the pandemic on their businesses (Figure 6).

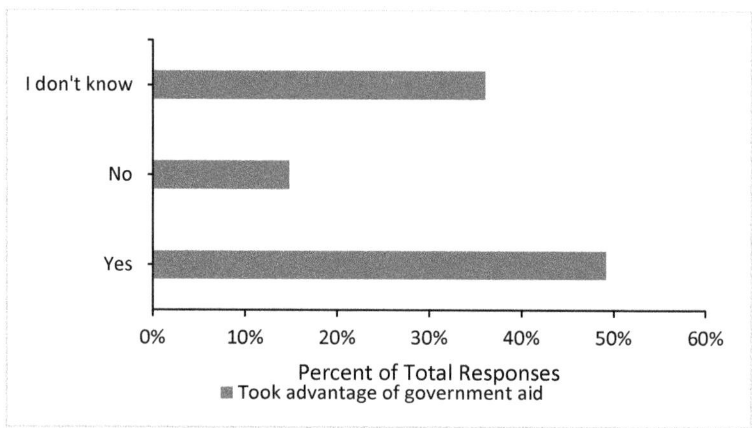

Figure 6. Participants' responses on the use of government aid to mitigate pandemic effects.

Table 3 shows the mean scores for the respondents' agreement with the statements under each category calculated based on the five-point Likert scale previously mentioned in the methodology section. Thus, averages closer to five indicate stronger agreement with the impact of COVID in the particular categories. Consequently, the participants reported the "communication" category as the most impacted by the COVID-19 pandemic, followed by the impact on the "construction projects," "company management," and "health and safety" categories.

Table 3. Descriptive statistics of statement categories.

Category	N	Mean	Std. Deviation
Company management	61	3.22	0.84
Construction projects	61	3.94	0.70
Health and safety	61	2.99	1.06
Communication	61	4.66	0.655

4.2. Comparison of Responses for Participants That Worked in New York City vs. Those That Worked in Other Parts of New York State

The data were analyzed using Mann–Whitney tests based on whether the participants worked in or out of New York City (Table 4). The participants who worked in New York City experienced significantly more reductions in their construction project teams. They also had more difficulty complying with the increased health and safety regulations and the new PPE requirements. The same group experienced higher COVID-19 positivity rates than those who worked in other parts of New York State. It is also important to note that fewer people in New York City reported that their companies embraced new technology during the pandemic compared to those in the other parts of New York State. The Mann–Whitney U test illustrated that those who worked in New York City were less likely to work remotely than those in other parts of New York State.

Table 4. Comparison of responses of those that worked in New York City vs. those who worked in the rest of New York State.

Statements/Question	Work in New York City	N	Mean Rank	Mann–Whitney U	Asymp. Sig. *
CM8: My company embraced new technology as a result of the pandemic	Yes	20	25.00	290.000	0.043 *
	No	41	33.93		
CP10: The pandemic reduced the size of my project team	Yes	20	37.92	271.500	0.029 *
	No	41	27.62		
HS20: COVID-19 Health and safety regulations were difficult to comply with	Yes	20	39.60	238.000	0.006 *
	No	41	26.80		
HS21: The new PPE requirements were difficult to comply with	Yes	20	38.33	263.500	0.021 *
	No	41	27.43		
HS22: There were high COVID-19 positivity rates on my projects	Yes	20	40.92	211.500	0.02 *
	No	41	26.16		
Ability to work from home ^	Yes	20	37.28	284.500	0.019 *
	No	41	27.94		

* Significant at a 95% confidence level; ^ question was on a 2-point scale.

4.3. Comparison of Responses from Builders vs. Non-Builders

Independent samples t-tests were conducted to identify significant differences in the pandemic's impact on builders (field-based) vs. non-builders (Table 5). The results illustrate that those in the builder category indicated significantly more contract performance issues. There was also a significantly higher agreement by the builders when asked if their companies requested extended project deadlines. In contrast, the non-builders agreed significantly more than the builders that their companies embraced newer technology due to the pandemic. Finally, the authors identified a significant difference between the builders and the non-builders regarding their ability to work from home during the pandemic. Builders were significantly less likely to work from home compared to the non-builders. However, there was no significant difference between the impacts of the pandemic on the productivity of the builders vs. non-builders.

Table 5. Comparison of responses of builders vs. non-builders.

Statements/Question	Builder	N	Mean	Std. Deviation	Std. Error Mean	Sig. *
CM1: My company experienced contract performances issues due to the pandemic	Yes	30	4.30	0.877	0.160	0.037 *
	No	31	3.68	1.351	0.243	
CM8: My company embraced new technology as a result of the pandemic	Yes	30	4.07	0.785	0.143	0.008 *
	No	31	4.58	0.672	0.121	
CP17: My company requested extended project deadlines due to the pandemic	Yes	30	4.37	0.669	0.122	0.043 *
	No	31	3.84	1.241	0.223	
Ability to work from home ^	Yes	30	1.50	0.509	0.093	0.012 *
	No	31	1.19	0.402	0.072	

* Significant at a 95% confidence level; ^ question was on a 2-point scale.

4.4. Survey Respondents' Recommendations for Potential Disruptions

A few participants responded to an optional open-ended question asking for any relevant information on the impact of COVID-19 on their companies and projects. Responses included comments on budget, workforce training, project schedules, and communication. It is also clear that the pandemic's effects may have differed based on the project type and whether projects were considered essential or not. The responses to another optional open-ended question asking for participants' recommendations for the construction industry to better prepare for future disruptions centered on the following topics:

4.4.1. Contract Language

The respondents indicated a need for clear contract terms and a Force Majeure clause in construction contracts especially related to project delays associated with a global pandemic. The proper documentation of responses to interruptions was suggested, which could be presented as evidence in the case of disputes or claims. The legal implications of COVID-19 in the construction industry have been discussed in other studies, and it is critical to ensure that contract documents are correctly interpreted and the teams on construction projects are aware of their roles in case of similar disruptions [54].

4.4.2. Supply Chain Management

The need for early notification to better prepare for supply chain interruptions and delays was discussed. This can be achieved through data-driven approaches, the Internet of Things (IoT), and data analytics to address possible supply-chain disruptions [55]. The construction industry needs to be more resilient and get better at real-time managing and monitoring of inventory through sophisticated systems, models, and technologies so that disruptions like this will not disarm the construction industry [56,57].

4.4.3. Project Planning and Scheduling

It was recommended to keep relevant stakeholders updated by ensuring clear communication about the status of projects and anticipated changes. Contingency planning was also suggested for materials and trades. The better utilization of construction project schedules through more input and approaches such as critical chain scheduling may help address the needs of different stakeholders as an impactful management tool [58]. Resource loading with the associated leveling of resources can be adopted, ensuring that the data are communicated and shared with relevant stakeholders, including an appropriate level of detail. Planning and management dashboards that analyze resource constraints also benefit construction projects. Overall, it is critical to understand project needs and how to prepare for possible disruptions adequately.

4.4.4. Remote Work Accommodations

The COVID-19 lockdown forced many individuals to work from home; fortunately, non-field-based construction-related tasks could be completed remotely. Several participants mentioned the need to embrace new technology to support remote work and virtual meetings. In line with the current literature, using innovative technologies to improve remote work productivity was beneficial to many, while some individuals needed additional support to overcome technical difficulties as they learned new technologies [23]. It is also important to note that remote work has led to social isolation, negatively impacting the mental health of workers [41]. One of the challenges of the construction industry is the need for in-person work since most field tasks cannot be completed remotely. Thus, there is still a need for more efficient and innovative solutions, including but not limited to technologies involving modular construction, IoT, drones, and 3D printing to improve productivity and reduce overall construction duration and personnel needed on the field [59,60].

4.4.5. Health and Safety Regulations and Guidance

The need to adhere to COVID-19 safety protocols was discussed to keep the workplace safe. One participant recommended that a task force should be formed to develop a central document that the construction industry can refer to for guidance. While the Occupational Safety and Health Administration (OSHA) published an advisory document to control and prevent exposure to COVID-19 for construction workers in 2020 [61], the publication is not yet a standard or regulation and lacks clarity on the circumstances under which it can be utilized. It is also important that construction safety personnel incorporate any current or future regulatory guidance in their company operational procedures, while the same is expected from site-specific safety plans developed for each construction project. Another suggestion from the survey responses called for more accessible on-site testing for COVID-19 infections. While access to on-site testing has significantly improved since the beginning of the pandemic, there is no central guidance or protocols for construction sites to increase the effectiveness, reliability, and safety of testing or other screening strategies such as temperature checks and employee questionnaires.

5. Discussion

The COVID-19 pandemic has caused significant disruptions to the construction industry, considering company management, construction projects, health and safety, and communication. Similar to other industries, the construction industry had to make significant changes in daily operations, such as mandating some non-field-based workers to work remotely, reducing the workforce on project sites, and utilizing virtual meetings. Participants reported having difficulty in obtaining new clients and projects. Based on the findings in this study, the professionals indicated the most significant impact on meeting modalities, and many companies seemed to embrace new telecommunication technology as a result of the pandemic; also, in-person meetings were limited, which is consistent with the findings from a study of construction professionals in the United States [40]. This could have been due to federal and state mandates during the nationwide lockdowns and the need to meet physical distancing requirements. However, field-based personnel classified as builders benefited significantly less from the remote work opportunities when compared to architects, engineers, and clients due to the nature of their tasks being more field-based. Virtual meetings were not always effective, as most people reported missing personal connections and non-verbal cues [40]. It was recommended that virtual communication should be improved with the use of reliable internet connectivity and effective technology [40].

Many participants did not report their construction firms having legal consequences on their projects due to the pandemic, but builders experienced more contract performance issues than non-builders. Legal consequences could emerge in the future resulting from the COVID-19 pandemic, as Alsharef et al. [23] reported that there would be a significant increase in the number of disputes, litigations, and claims in the construction industry. Therefore, the recommendations to strengthen contract terms and keep proper records would be beneficial. Similarly, Chivilo, Fonte, and Koger [17] suggest providing notices to contractual partners and ensuring the timely exchange of information between all parties. Furthermore, the authors identified non-builders and those who work primarily outside of New York City as more likely to state that their companies embraced new technology as a result of the pandemic. Almost half of the respondents reported that their company took advantage of government aid to mitigate the effects of the pandemic on their businesses. Similarly, Brown, Brooks, and Dong [18] noted that 64% of US small businesses in construction got financial support from the paycheck protection program, while 61% of non-farm industries received the same, so financial assistance was instrumental in offsetting overhead costs in the construction industry.

In their study, Gamil and Alhagar [29] found that the most significant impacts of the pandemic on the construction industry were the suspension of projects, labor impact and job loss, time overruns, cost overruns, and financial implications. In terms of the impact of the pandemic on construction projects, many projects experienced schedule interruptions and delays, which is a result consistent with the findings of Morris [25] and Ogunnusi et al. [26]. While almost all stakeholders involved in the construction industry experienced scheduling delays and material supply issues due to the pandemic [23], our study revealed that builders were more likely to state having contractual issues and extended project schedules. Although the current literature illustrates an overall loss of productivity in the construction industry [23], those who worked from home reported that their productivity increased during the pandemic. It is also important to note that there was no significant difference between the responses from builders and non-builders regarding their productivity; however, builders were less likely to work from home than non-builders since their work is mainly field-based.

Regarding health and safety on construction projects, the pandemic seems to have been managed well on most construction projects, as very few people reported high COVID-19 positivity rates on their projects, but participants in New York City reported significantly more staff reductions due to higher positivity rates than those in other parts of New York State. Certain PPE requirements and safety protocols were already in place for construction projects prior to the pandemic, and some tasks required the use of nose masks and goggles for safety [37]. Limiting the number of people on project sites also contributed to safer working conditions. The higher rates of COVID-19 reported by those in New York City reflect the statewide rates, especially since New York City was the epicenter of the spread of COVID-19 in the US in the early days of the pandemic in 2020 [13]. Furthermore, those in New York City found it more challenging to comply with COVID-19 health and safety regulations and the new PPE requirements than those in other parts of New York State. It is important to note that this study did not investigate whether the regulations differed in New York City from the rest of New York State. Yet, these findings are consistent with the observations of Bushman et al. [45] from the case report of two construction sites in New York City, emphasizing the construction industry's challenges in complying with health and safety guidelines. The widespread availability of COVID-19 vaccines has brought a glimmer of hope to reduce deaths and hospitalization as individuals and nations work towards recovering from the negative impacts of the pandemic.

6. Conclusions

As industries begin to recover from the effects of the pandemic, effective strategies should be put in place to aid recovery. The findings of this study relate to the pandemic's impact on the construction industry in New York State based on four themes, namely company management, construction projects, health and safety, and communication. This study also explored the impact of the pandemic on builders and non-builders and those working within and outside New York City. Overall, most non-essential construction projects suffered cost and time overruns during the pandemic. The results of this study indicate that the most impacted category was communication, followed by construction projects, company management, and health and safety. In some respects, the pandemic had varying impacts on builders and non-builders and those working in and out of New York City. Therefore, the industry needs to consider such differences as it addresses current and future similar challenges facing the construction industry.

The results of this study highlight several areas of improvement for the construction industry to be more resilient and better prepared for similar future disruptions. These areas include contract language, supply-chain management strategies, project planning and scheduling, remote work accommodations, virtual communication techniques and technology, and health and safety regulations and guidance. Some of the lessons learned in these areas include the need for companies to be diligent with their contract documents and language to ensure clear communication between relevant stakeholders. For companies that choose to primarily utilize virtual communication approaches, it is critical to consider potential connectivity issues and ensure that their platforms are reliable. To ensure the safety of construction workers, robust health and safety guidelines should be followed. Regulatory bodies may need to address such needs to provide guidance and ensure consistency across the industry. There is also a clear need for collaboration among businesses and public authorities to ensure construction workers understand the importance of complying with the guidelines.

As the pandemic wanes or potentially shifts toward an endemic, further studies are needed to determine the true extent of its impact on the construction industry and the daily activities on construction sites in New York State and globally. Such studies would focus on the extent of unexpected disruptions and how the impacts differ among various construction industry sectors and project types. The study's main limitations include the relatively small sample size, which may impact the generalizability of the findings and introduce bias since a non-probability sampling method was used, and the fact that those who chose to participate may not be representative of the entire population of individuals in the building industry in New York State. In addition, approximately 90% of the respondents in this study were employed by private companies, suggesting future research may also focus on the impacts of the pandemic on public agencies that play an essential role in the construction industry in New York State. Finally, this study did not capture the long-term impacts of the pandemic on the construction industry in New York State since such conclusions may require longitudinal studies spanning several years.

Author Contributions: Conceptualization, E.I. and Y.S.A.; methodology, Y.S.A. and B.G.C.; validation, Y.S.A. and B.G.C.; formal analysis, E.I., Y.S.A. and B.G.C..; investigation, E.I.; data curation, E.I.; writing—original draft preparation, E.I., Y.S.A. and B.G.C.; writing—review and editing, Y.S.A. and B.G.C.; Supervision, Y.S.A. All authors have read and agreed to the published version of the manuscript.

Funding: This research received no external funding.

Institutional Review Board Statement: The study was conducted in accordance with the Declaration of Helsinki and approved by the Institutional Review Board (or Ethics Committee) of Rochester Institute of Technology (protocol code 02021621 and date of approval- 26 February 2021).

Informed Consent Statement: Informed consent was waived for all subjects involved in the study due to the study being exempt.

Data Availability Statement: Data supporting reported results are available from the corresponding authors on request.

Conflicts of Interest: The authors declare no conflict of interest.

References

1. Boskin, M.J. *How Does the COVID Recession Compare?* World Economic Forum: Cologny, Switzerland, 2020. Available online: https://www.weforum.org/agenda/2020/08/how-does-the-covid-recession-compare/ (accessed on 1 September 2021).
2. Shen, H.; Fu, M.; Pan, H.; Yu, Z.; Chen, Y. The impact of the COVID-19 pandemic on firm performance. *Emerg. Mark. Financ. Trade* **2020**, *56*, 2213–2230. [CrossRef]
3. Ozili, P.K.; Arun, T. Spillover of COVID-19: Impact on the Global Economy. 2020. Available online: https://papers.ssrn.com/sol3/papers.cfm?abstract_id=3562570 (accessed on 1 May 2020).
4. Bachman, D. The Economic Impact of COVID-19 (Novel Coronavirus). Deloitte Insights. Available online: https://www2.deloitte.com/us/en/insights/economy/covid-19/economic-impact-covid-19.html (accessed on 1 September 2021).
5. Maliszewska, M.; Mattoo, A.; van der Mensbrugghe, D. *The Potential Impact of COVID-19 on GDP and Trade: A Preliminary Assessment*; World Bank Policy Research Working Paper no. 9211; World Bank Group: Washington, DC, USA, 2020.
6. Kurt, D. The Special Economic Impact of Pandemics. Investopedia. Available online: https://www.investopedia.com/special-economic-impact-of-pandemics-4800597 (accessed on 1 September 2021).
7. de Best, R. Value Added of U.S. Construction Industry as a Percentage of GDP 2007–2020. Statista. Available online: https://www.statista.com/statistics/192049/value-added-by-us-construction-as-a-percentage-of-gdp-since-2007/ (accessed on 1 September 2021).
8. Simonson, K. The Economic Impact of Construction in the United States and New York. AGC. Available online: https://www.agc.org/sites/default/files/Files/Construction%20Data/NY.pdf (accessed on 20 July 2021).
9. McKinsey Global Institute. *Reinventing Construction: A Route of Higher Productivity*; McKinsey Company: New York, NY, USA, 2017.
10. International Labor Organization. *Impact of COVID-19 on the Construction Sector*; International Labor Organization: Geneva, Switzerland, 2021. Available online: https://www.ilo.org/wcmsp5/groups/public/---ed_dialogue/---sector/documents/briefingnote/wcms_767303.pdf (accessed on 1 September 2021).
11. Haydon, D.; Kumar, N.; Bloom, J. *Industries Most and Least Impacted by COVID-19 from a Probability of Default Perspective*; S&P Global Market Intelligence: New York, NY, USA, 2021.
12. The New York Times. The New York Times. New York City Coronavirus Map and Case Count. *The New York Times*, 15 December 2021.
13. Thompson, C.N.; Baumgartner, J.; Pichardo, C.; Toro, B.; Li, L.; Arciuolo, R.; Chan, P.Y.; Chen, J.; Culp, G.; Davidson, A.; et al. COVID-19 Outbreak—New York City, 29 February–1 June 2020. *Morb. Mortal. Wkly. Rep.* **2020**, *69*, 1725. [CrossRef] [PubMed]

14. OSC. *The Construction Industry in New York City: Recent Trends and the Impact of COVID-19*; Office of the New York State Comptroller: New York, NY, USA, 2021; Available online: https://www.osc.state.ny.us/files/reports/osdc/pdf/report-3-2021.pdf (accessed on 2 February 2022).

15. Bousquin, J. *Construction Accounts for the Most COVID Deaths of Any Industry in Colorado*; Construction Dive: Washington, DC, USA, 2021. Available online: https://www.constructiondive.com/news/construction-accounts-for-the-most-covid-19-deaths-of-any-industry-in-color/604705/ (accessed on 25 August 2021).

16. IMF. *World Economic Outlook: A Long and Difficult Ascent*; International Monetary Fund: Washington, DC, USA, 2020.

17. Chivilo, J.P.; Fonte, G.A.; Koger, G.H. *A Look at COVID-19 Impacts on the Construction Industry*; Holland & Knight: Tampa, FL, USA, 2020. Available online: https://www.hklaw.com/en/insights/publications/2020/05/a-look-at-covid19-impacts-on-the-construction-industry (accessed on 25 August 2021).

18. Brown, S.; Brooks, R.D.; Dong, X.S. *Impact of COVID-19 on Construction Workers and Businesses*; CDC: Atlanta, GA, USA, 2020.

19. Krebs, C. *Advisory Memorandum on Identification of Essential Critical Infrastructure Workers during COVID-19 Response*; CISA: Rosslyn, VA, USA, 2020.

20. Empire State Development. *Guidance for Determining Whether a Business Enterprise is Subject to a Workforce Reduction under Recent Executive Orders*; New York State: New York, NY, USA, 2020. Available online: https://esd.ny.gov/guidance-executive-order-2026 (accessed on 1 September 2021).

21. Layton, D. *The Extraordinary and Unexpected Pandemic Increase in House Prices: Causes and Implications*; Havard Joint Center for Housing Studies: Cambridge, MA, USA, 2021. Available online: https://www.jchs.harvard.edu/blog/extraordinary-and-unexpected-pandemic-increase-house-prices-causes-and-implications (accessed on 1 August 2021).

22. Johnson, P.; Reizen, R. *COVID-19, A New President and Economic Uncertainty: What to Expect in the Construction Industry in 2021*; Gould & Ratner LLP: Chicago, IL, USA, 2021; Available online: https://www.jdsupra.com/legalnews/covid-19-a-new-president-and-economic-7621707/ (accessed on 1 September 2021).

23. Alsharef, A.; Banerjee, S.; Uddin, S.M.; Albert, A.; Jaselskis, E. Early impacts of the COVID-19 pandemic on the United States construction industry. *Int. J. Environ. Res. Public Health* **2021**, *18*, 1559. [CrossRef] [PubMed]

24. Choi, S.D.; Staley, J. Safety and Health Implications of COVID-19 on the United States Construction Industry. *Ind. Syst. Eng. Rev.* **2021**, *9*, 56–67. [CrossRef]

25. Morris, G.D. *6 Critical COVID-19 Risks for the Construction Industry*; Risk and Insurance: Blue Bell, PA, USA, 2020. Available online: https://riskandinsurance.com/6-critical-covid-19-risks-for-the-construction-industry/ (accessed on 1 September 2021).

26. Ogunnusi, M.; Hamma-Adama, M.; Salman, H.; Kouider, T. COVID-19 pandemic: The effects and prospects in the construction industry. *Int. J. Real Estate Stud.* **2020**, *14*, 120–128.

27. AGC. *2021 Workforce Survey Results: New York Results*; AGC and AUTODESK: New York, NY, USA, 2021. Available online: https://www.agc.org/sites/default/files/2021_Workforce_Survey_NY.pdf (accessed on 7 September 2021).

28. Banoobhai, J.; Osman, N. COVID to Impact Construction Claims 'Until at Least 2023'. Pinsent Masons, 2021. Available online: https://www.pinsentmasons.com/out-law/analysis/covid-impact-construction-claims-2023 (accessed on 30 August 2021).

29. Gamil, Y.; Alhagar, A. The impact of pandemic crisis on the survival of construction industry: A case of COVID-19. *Mediterr. J. Soc. Sci.* **2020**, *11*, 122. [CrossRef]

30. Alenezi, T.A.N. Covid-19 Causes of Delays on Construction Projects in Kuwait. *Int. J. Eng.* **2020**, *8*, 35–39.

31. Rubin, D.K. *COVID-19 Project Impact Liability Is Confused, Says Consultant Survey*; ENR: New York, NY, USA, 2021. Available online: https://www.enr.com/articles/51200-covid-19-project-impact-liability-is-confused-says-consultant-survey (accessed on 1 September 2021).

32. AGC. *2021 Workforce Survey Results: National Results*; AGC and AUTODESK: Arlington, VA, USA, 2021. Available online: https://www.agc.org/sites/default/files/2021_Workforce_Survey_National_Autodesk.pdf (accessed on 7 September 2021).

33. Biörck, J.; Sjödin, E.; Blanco, J.L.; Mischke, J.; Strube, G.; Ribeirinho, M.J. *How Construction Can Emerge Stronger after Coronavirus*; McKinsey & Company: New York, NY, USA, 2020.

34. Assaad, R.; El-adaway, I.H. Guidelines for Responding to COVID-19 Pandemic: Best Practices, Impacts, and Future Research Directions. *J. Manag. Eng.* **2021**, *37*, 06021001. [CrossRef]

35. Pinheiro, M.D.; Luís, N.C. COVID-19 could leverage a sustainable built environment. *Sustainability* **2020**, *12*, 5863. [CrossRef]

36. Pasco, R.F.; Fox, S.J.; Johnston, S.C.; Pignone, M.; Meyers, L.A. Estimated association of construction work with risks of COVID-19 infection and hospitalization in Texas. *JAMA Netw. Open* **2020**, *3*, e2026373. [CrossRef]

37. Stiles, S.; Golightly, D.; Ryan, B. Impact of COVID-19 on health and safety in the construction sector. *Hum. Factors Ergon. Manuf. Serv. Ind.* **2021**, *31*, 425–437. [CrossRef]

38. New York State. *Reopening New York: Construction Guidelines for Employers and Employees*; New York State: New York, NY, USA, 2020. Available online: https://www.zdlaw.com/assets/htmldocuments/NYConstructionShortGuidelines.pdf (accessed on 1 September 2021).

39. Labs, O. State of Remote Work. 2021. Available online: https://resources.owllabs.com/state-of-remote-work/2020 (accessed on 17 October 2021).

40. Encinas, E.; Simons, A.; Sattineni, A. Impact of COVID-19 on Communications within the Construction Industry. *EPiC Ser. Built Environ.* **2021**, *2*, 165–172.

41. Pamidimukkala, A.; Kermanshachi, S.; Nipa, T.J. Impacts of COVID-19 on Health and Safety of Workforce in Construction Industry. In Proceedings of the International Conference on Transportation and Development 2021, Virtual Conference, 8–10 June 2021; pp. 418–430.
42. Brynjolfsson, E.; Horton, J.J.; Ozimek, A.; Rock, D.; Sharma, G.; TuYe, H.-Y. *COVID-19 and Remote Work: An Early Look at US Data*; National Bureau of Economic Research: Cambridge, MA, USA, 2020.
43. Robinson, B. *Is Working Remote a Blessing or Burden? Weighing the Pros and Cons*; Forbes: Jersey City, NJ, USA, 2020. Available online: https://www.forbes.com/sites/bryanrobinson/2020/06/19/is-working-remote-a-blessing-or-burden-weighing-the-pros-and-cons/?sh=34ad478440a9 (accessed on 30 May 2021).
44. Kaushik, M.; Guleria, N. The impact of pandemic COVID-19 in workplace. *Eur. J. Bus. Manag.* **2020**, *12*, 9–18.
45. Bushman, D.; Sekaran, J.; Jeffery, N.; Rath, C.; Ackelsberg, J.; Weiss, D.; Wu, W.; van Oss, K.; Johnston, K.; Huang, J.; et al. Coronavirus Disease 2019 (COVID-19) Outbreaks at 2 Construction Sites—New York City, October–November 2020. *Clin. Infect. Dis.* **2021**, *73*, S81–S83. [CrossRef]
46. Setia, M.S. Methodology series module 3: Cross-sectional studies. *Indian J. Dermatol.* **2016**, *61*, 261. [CrossRef]
47. Lavrakas, P.J. *Encyclopedia of Survey Research Methods (Vols. 1-0)*; Sage Publications: New York, NY, USA, 2008. Available online: https://methods.sagepub.com/reference/encyclopedia-of-survey-research-methods/n120.xml (accessed on 5 September 2021).
48. Nemoto, T.; Beglar, D. Likert-Scale Questionnaires. In *JALT 2013 Conference Proceedings*; JALT: Tokyo, Japan, 2014; pp. 1–8.
49. Losby, J.; Wetmore, A. *Using Likert Scales in Evaluation Survey Work*; CDC: Atlanta, GA, USA, 2012.
50. Etikan, I.; Musa, S.A.; Alkassim, R.S. Comparison of convenience sampling and purposive sampling. *Am. J. Theor. Appl. Stat.* **2016**, *5*, 1–4. [CrossRef]
51. Ott, R.L.; Longnecker, M.T. *An Introduction to Statistical Methods and Data Analysis*; Cengage Learning: Boston, MA, USA, 2015.
52. Zamanzadeh, V.; Ghahramanian, A.; Rassouli, M.; Abbaszadeh, A.; Alavi-Majd, H.; Nikanfar, A.-R. Design and implementation content validity study: Development of an instrument for measuring patient-centered communication. *J. Caring Sci.* **2015**, *4*, 165. [CrossRef]
53. Siegle, D. *Instrument Reliability*; NEAG School of Education, University of Connecticut: Storrs, CT, USA, 2002. Available online: https://researchbasics.education.uconn.edu/instrument_reliability/# (accessed on 15 May 2021).
54. MAl-Mhdawi, K.S.; Brito, M.P.; Nabi, M.A.; El-Adaway, I.H.; Onggo, B.S. Capturing the impact of COVID-19 on construction projects in developing countries: A case study of Iraq. *J. Manag. Eng.* **2022**, *38*, 05021015. [CrossRef]
55. Dallasega, P.; Rauch, E.; Linder, C. Industry 4.0 as an enabler of proximity for construction supply chains: A systematic literature review. *Comput. Ind.* **2018**, *99*, 205–225. [CrossRef]
56. Ivanov, D.; Dolgui, A. Viability of intertwined supply networks: Extending the supply chain resilience angles towards survivability. A position paper motivated by COVID-19 outbreak. *Int. J. Prod. Res.* **2020**, *58*, 2904–2915. [CrossRef]
57. Christopher, M.; Peck, H. *Building the Resilient Supply Chain*; Cranfield University: Cranfield, UK, 2004.
58. Steyn, H. Project management applications of the theory of constraints beyond critical chain scheduling. *Int. J. Proj. Manag.* **2002**, *20*, 75–80. [CrossRef]
59. Innella, F.; Arashpour, M.; Bai, Y. Lean methodologies and techniques for modular construction: Chronological and critical review. *J. Constr. Eng. Manag.* **2019**, *145*, 04019076. [CrossRef]
60. Oesterreich, T.D.; Teuteberg, F. Understanding the implications of digitisation and automation in the context of Industry 4.0: A triangulation approach and elements of a research agenda for the construction industry. *Comput. Ind.* **2016**, *83*, 121–139. [CrossRef]
61. Occupational Safety and Health Administration. *COVID-19: Control and Prevention—Construction Work*; Occupational Safety and Health Administration: Washington, DC, USA, 2020.

Article

Spatial and Temporal Analysis of Extreme Precipitation under Climate Change over Gandaki Province, Nepal

Sudip Pandey * and Binaya Kumar Mishra

School of Engineering, Pokhara University, Pokhara 33700, Nepal
* Correspondence: er.sudippandey@gmail.com

Abstract: This paper presents a research study of expected precipitation extremes across the Gandaki Province, Nepal. The study used five indices to assess extreme precipitation under climate change. Precipitation output of two Global Climate Models (GCMs) of Coupled Model Intercomparison Project Phase Six (CMIP6) were used to characterize the future precipitation extremes during the rainfall season from June to September (JJAS) and overall days of the year. To characterize extreme precipitation events, we used daily precipitation under the SSP2–4.5 and SSP5–8.5 scenarios from the Beijing Climate Center and China Meteorological Administration, China; and Meteorological Research Institute (MRI), Japan. Considering large uncertainties with GCM outputs and different downscaling (including bias correction) methods, direct use of GCM outputs were made to find change in precipitation pattern for future climate. For 5-, 10-, 20-, 50-, and 100-year return periods, observed and projected 24 h and 72 h annual maximum time series were used to calculate the return level. The result showed an increase in simple daily intensity index (SDII) in the near future (2021–2040) and far future (2081–2100), with respect to the base-year (1995–2014). Similarly, heavy precipitation days (R50 mm), very heavy precipitation days (R100 mm), annual daily maximum precipitation (RX1day), and annual three-day maximum precipitation (RX3day) indices demonstrated an increase in extreme precipitation toward the end of the 21st century. A comparison of R50 mm and R100 mm values showed an extensive (22.6% and 63.8%) increase in extreme precipitation days in the near future and far future. Excessive precipitation was forecasted over Kaski, Nawalparasi East, Syangja, and the western half of the Tanahun region. The expected increase in extreme precipitation may pose a severe threat to the long-term viability of social infrastructure, as well as environmental health. The findings of these studies will provide an opportunity to better understand the origins of severe events and the ability of CMIP6 model outputs to estimate anticipated changes. More research into the underlying physical factors that modulate the occurrence of extreme incidences expected for relevant policies is suggested.

Keywords: climate change; CMIP6; extreme precipitation; Gandaki Province; GIS

Citation: Pandey, S.; Mishra, B.K. Spatial and Temporal Analysis of Extreme Precipitation under Climate Change over Gandaki Province, Nepal. *Architecture* **2022**, *2*, 724–759. https://doi.org/10.3390/architecture2040039

Academic Editors: Iftekhar Ahmed, Masa Noguchi, David O'Brien, Chris Tucker and Mittul Vahanvati

Received: 28 March 2022
Accepted: 14 November 2022
Published: 22 November 2022

1. Introduction

Climate change is generally considered to be increasing warming days, erratic rainfall patterns, ecological variability, biological change, and the adverse effects these things have on human beings. It is measured by finding the patterns of deviation in temperature, humidity, atmospheric pressure, wind, precipitation, atmospheric particle counts, and other meteorological variables in a specific region over a long period [1]. Different research studies show that temperature increases rapidly, the rainfall pattern is erratic, seasons are changing, events of droughts are increasing, and hailstorm occurs abnormally [2]. The Intergovernmental Panel on Climate Change (IPCC) forecasts an increase in mean temperature of around 1 °C by 2025 and 3 °C by the end of the next century [3]. Climate change is one of the most critical global challenges of the century. It is expected that climate change will have a significant impact on extreme precipitation because the warming atmosphere will alter the precipitation pattern by changing the global/regional hydrological cycle.

Extreme precipitation is one of the major factors that trigger natural disasters, such as floods, landslides, and droughts, which cause the loss of life and property. Under climate change, rainfall is increasing its pattern in the world [4]. It shows the problems of food and water security, as well as disaster management. In 14–17 June 2013, the cloudburst happened in the northwestern region of Nepal and India border, killing around 5700 and affecting more than 100,000 people from both countries. Another event of 14–16 August 2014 caused massive flooding and triggered a number of landslides, causing massive losses of life and property and distressing around 35,000 households [4].

Mainly Tropical Disturbance, i.e., EI Nino-Southern oscillation and Madden Julian Oscillation, contributes to the variability of extreme rainfall in the South Pacific region. In tropical regions, tropical cyclones play the vital role for the extreme weather [5]. Precipitation in Nepal varies significantly from place to place due to sharp topographical variation. According to the rain-bearing wind approach, most rainfall occurred over the foothills of the lower Himalayas in monsoon season and least on the leeward side of individual successive range [6].

Climate change is one of the most critical challenges of the century for Nepal. The warming trend in Nepal after 1997 has ranged from 0.06 to 0.12 °C per year in most of the middle mountains and Himalayan regions [7]. Nepal is unique geographic location, with the Churia Range at the south and the Himalayas to the north, and the prevailing monsoon has made it one of the wettest countries of the world. It is located at the southern slope of the Central Himalayas, with latitudes of 26°22′ to 30°27″ north and longitudes of 80°04′ to 88°12′ east. The topography diverges from 60 m in south to the highest peak of the world, i.e., 8848 m. Pre-monsoon (March–May), monsoon (June–September), post-monsoon (October–November), and winter (December–February) are the four main seasons of Nepal. The majority of the precipitation occurs during the monsoon season, which accounts for greater than 80% of the annual precipitation.

Predictions are constantly being made to understand what will happen based on observed climate shifts over the past decades and projected greenhouse gas emissions. Information on the likely change in the precipitation pattern is highly useful in various fields, such as engineering, environment, forestry, and agriculture. Investigations on climate-change impacts on precipitation extremes are necessary for proper engineering designs of hydrologic structures such as dams, bridges, irrigation canals, sewerage, etc. Rainfall intensity–duration–frequency (IDF) curves are used to quantify the depth and its frequency of rainfall, thinking about one-of-a-kind periods; historically, the IDF curves are primarily based totally on historic rainfall residences in general [8].

Global Climate Model (GCM) is a popular tool for generating the future climate-change scenarios. Recently, Coupled Model Intercomparison Project Phase Six (CMIP6) simulations have been used to find the change pattern in future precipitation extremes. GCM faces problems of reliable precipitation simulation at regional or local scales, particularly in regard to the simulation of extreme precipitation events, due to its coarse spatial resolution and the current incomplete understanding of the climate system. The poor performance in regional/local precipitation simulation makes it difficult to directly use GCM outputs in studies on climate-change impact on extreme precipitation change, because an extreme precipitation event is most likely a localized phenomenon. Recently, research efforts have been made to downscale the GCM to support localized impact assessment. However, such downscaling models are either complex and computationally demanding (dynamic downscaling) or require the extensive use of observed data (statistical downscaling). That leads to a limited availability of model results, which, in most cases, are not sufficient to fulfil localized impact-assessment needs [9].

Precipitation-change assessments for Nepal at different scales (grid cell, basin, district, and country) and the temporal scale can be found in different forms (reports and journal papers). However, there is no study on climate-change impact assessment primarily focused on precipitation extremes over Gandaki Province. This research investigated implications of climate change on characteristics of extreme precipitation over Gandaki

Province of Nepal. Gandaki is one of the seven provinces (states) of Nepal. Using CMIP6 simulations, this research investigated likely change in precipitation extremes over Gandaki Province, Nepal. To assess extreme precipitation, the study used five indices: simple daily intensity index (SDII), heavy precipitation days (R50 mm), very heavy precipitation days (R100 mm), annual daily maximum precipitation (RX1day), and annual three-day maximum precipitation (RX3day).

2. Study Area and Methods

In order to understand the precipitation pattern over the past decades, the observed data of the previous year were obtained through the relevant department. The rainfall data were collected from the Department of Hydrology and Metrology, Nepal. GCM precipitation outputs for different scenarios were obtained online. An analysis of extreme precipitation under climate-change consequence was conducted through an MS Excel-based tool. The overall methodology and how it relates to each objective are briefly shown below, in diagrammatic form (Figure 1).

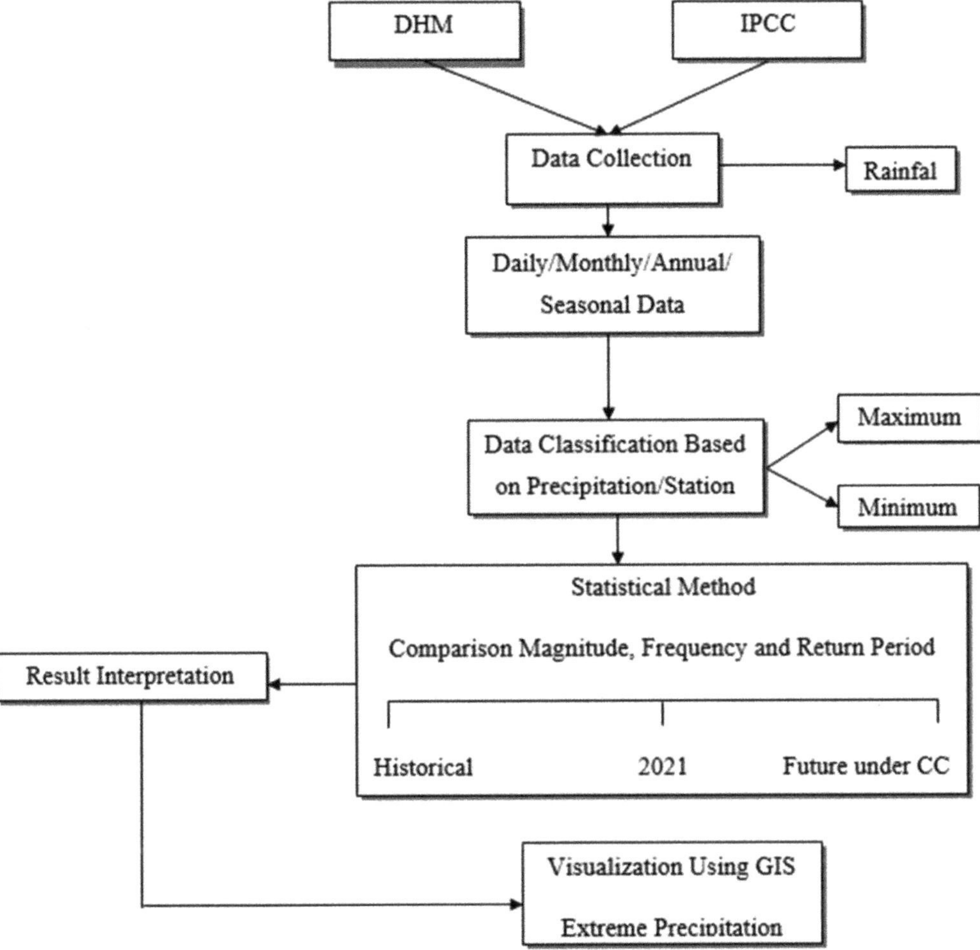

Figure 1. Conceptual framework for this study.

2.1. Study Area

The location of the study area is Gandaki Province, Nepal (Figure 2a). It is one of the seven federal provinces established by the current constitution of Nepal, which was promulgated on 20 September 2015 AD. There are 11 districts: Manang, Gorkha, Lamjung, Tanahun, Nawalparasi (Bardaghat-Susta East), Kaski, Syangja, Parbat, Baglung, Magdi, and Mustang. It borders the Tibet Autonomous Region of China to the north, Bagmati Province to the east, Karnali Province to the west, and Lumbini Province and Bihar of India to the south. It has an area of 21,773 Km^2, which is about 14.66% of the total area of Nepal. The state lies at a latitude of 27°20′ N~29°20′ N and longitude of 82° 52′ E~85°12′ E [10]. In terms of terrain, the province lies in the Himalayan region, along with hilly and plain regions with high spatial variability of climate (Figure 2b). The summer monsoon, which lasts from June to September, has the most influence on the climate in the study area. Basically, the climate ranges from subtropical in the lower belt to arctic in the upper elevations [11].

Climate change is a major environmental challenge that plays out through changing intensity, duration, and frequency of extreme events in the Gandaki Province. Water-related disasters such floods and landslides are frequently reported during monsoon due to occurrences of heavy rainfall events. The Gandaki Province has a wide spatial heterogeneity in topography and climate, with the topographical variation ranging. Elevations range from 60 m in the south to higher than 8000 m in the north, where the watershed contains the Dhaulagiri (8167 m) and Annapurna (8091 m) peaks. The average annual precipitation varies from as low as 150 mm in the Trans-Himalayas to as high as 5400 mm in the hilly region. The Gandaki River Basin is the main river system. The presence of the Langtang, Machapuchare, Dhaulagiri, and Annapurna mountains have created a steep variation in climate over the basin.

(a)

Figure 2. *Cont.*

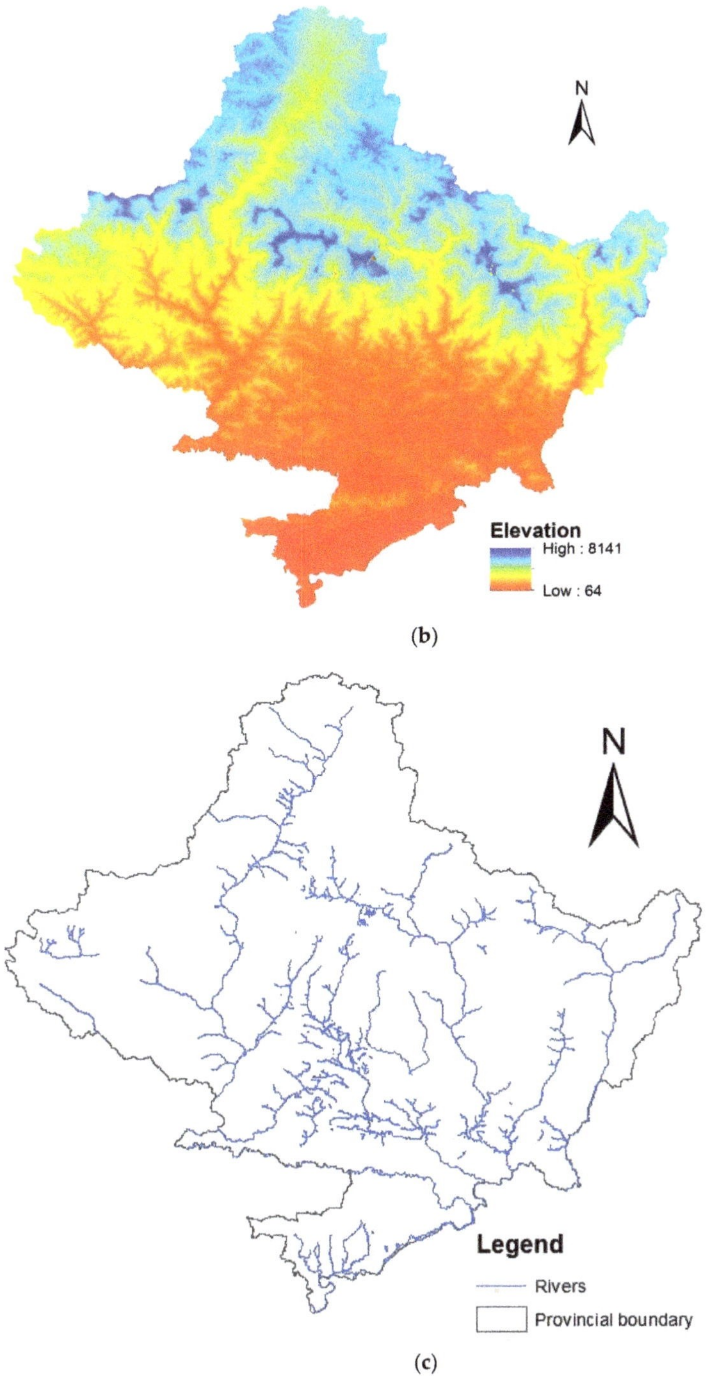

Figure 2. (**a**) Map of Nepal, showing Gandaki Province. (**b**) Elevation variation across Gandaki Province. (**c**) River network inside Gandaki Province.

2.2. Precipitation Data

2.2.1. Observed

The study was based upon meteorological data collected from Nepal's Department of Hydrology and Meteorology (DHM). The DHM is the country's only institution in charge of collecting and disseminating meteorological and hydrological data. It builds and maintains a nationwide network of hydrometeorological stations. DHM collects daily meteorological data, and these data of Gandaki Province were used for this research, as illustrated in (Table 1).

Table 1. List of meteorological stations used in this study.

SN	Station ID	Station Name	District	Latitude	Longitude	Altitude (m)	Period	Types of Station
1	601	Jomsom	Mustang	28°47′	8°343′	2744	2003–2020	Climatology
2	605	Baglung	Baglung	28°16′	83°36′	984	2003–2020	Climatology
3	606	Tatopani	Myagdi	28°29′	83°39′	1243	2003–2020	Precipitation
4	614	Kushma	Parbat	28°13′	83°42′	891	2003–2020	Climatology
5	615	Bobang	Baglung	28°24′	83°06′	2273	2003–2020	Precipitation
6	704	Beluwa/Girwari	Nawalpur	27°41′	84°03′	150	2003–2020	Precipitation
7	706	Dumkauli	Nawalpur	27°41′	84°13′	154	2003–2020	Agrometrology
8	802	Khudi Bazar	Lamjung	28°22′	84°54′	1334	2003–2020	Climatology
9	804	Pokhara Airport	Kaski	28°13′	84°00′	827	1995–2020	Aeronautical
10	805	Syangja	Syangja	28°06′	83°53′	868	2003–2020	Climatology
11	808	Bandipur	Tanahun	27°56′	84°25′	965	2003–2020	Climatology
12	809	Gorkha	Gorkha	28°00′	84°37′	1097	2003–2020	Agrometrology
13	814	Lumle	Kaski	28°18′	83°48′	1740	1995–2020	Agrometrology
14	817	Damauli	Tanahun	27°58′	84°17′	358	2003–2020	Climatology
15	820	Manang Bhot	Manang	28°40′	84°01′	3420	2003–2020	Climatology

Source: Department of Hydrology and Meteorology, 2021.

2.2.2. GCM

GCM outputs are used for climate predictions and to study climate variability and change. Most of the GCMs are characterized by a coarse grid resolution, resulting in greater uncertainty. Climate-change impact assessment based on GCMs is simulated by the integration of future emission paths (socioeconomic scenarios). There is a wide selection of climate models available to provide scenarios of future climate change. For meaningful climate-change-impact studies, realistic simulations of the local climate are needed. The use of multiple GCMs and scenarios is recommended due to the large uncertainty of the GCM outputs. However, the use of too many GCMs requires too much time and too many resources. In this case, two GCM datasets were used from the CMIP6 models. They are from the Beijing Climate Center and China Meteorological Administration, China (BCC-CSM2-MR); and the Meteorological Research Institute (MRI), Japan (Table 2). BCC-CSM2-MR outputs were considered because of the proximity of the study area to China, where this GCM was originated. On the other hand, MRI GCM outputs have been widely applied, as well as tested, in the Nepalese basin by the authors of previous works [12,13].

Table 2. Information of the CMIP6 climate models used in this study.

SN	Models	Institution	Resolution
1	BCC-CSM2-MR	Beijing Climate Center and China Meteorological Administration, China	1.13° × 1.13°
2	MRI-ESM2-0	Meteorological Research Institute (MRI), Japan	1.13° × 1.13°

The daily precipitation under the SSP2–4.5 and SSP5–8.5 scenarios was used to estimate extreme precipitation events [14]. The SSP2–4.5 (+4.5 Wm^2) pathway represents a medium-forcing middle-of-the-road pathway, whereas the SSP5–8.5 (+8.5 Wm^2) pathway represents a high-end-forcing pathway. The SSP2–4.5 scenario is thought to be a more likely conclusion, with modest mitigation efforts limiting global warming to a maximum of 2.5 degrees Celsius above pre-industrial levels by the end of the twenty-first century [14]. SSP5–8.5, on the other hand, is also known as "business as usual", meaning a dismal future that is fossil-fuel intensive and devoid of any climate mitigation, resulting in approximately 5 °C of warming by the end of the century. In this study, we considered 1995–2014 data as the baseline, as well as two future time periods, referred to as the near future (2021–2040) and far future (2081–2100), respectively.

2.3. Precipitation Indices

To investigate projected variations in precipitation over the research region, five indicators collected from ETCCDI were used. The indicators stated below, in Table 3, were chosen to investigate future changes in regional precipitation intensity, frequency, and duration [14]. A recent study on EA used the indices described above to see if CMIP6 models could accurately represent the current characteristics and geographical distribution of extreme precipitation. The study's key findings proved MME's capacity to describe observed precipitation and related extremes during rainfall seasons when compared to individual models. Furthermore, when compared to the extremes, MME performed better in recreating yearly total precipitation, according to the study. As a result of Reference [15], the MME was also used in this study to look at future changes in local precipitation extremes. The five indices shown in (Table 3) were used in numerous studies to analyze changes in precipitation over the Gandaki area.

Table 3. Extreme climate indices used in studies.

Label	Index Name	Index Definition	Unit
R50 mm	Heavy precipitation days	Number of days where PRij 1 > 50 mm of a given period	Days
R100 mm	Very heavy precipitation days	Number of days where PRij 1 > 100 mm of a given period	Days
RX1day	Max 1-day precipitation	RX1dayj = max (PRij), the maximum 1-day precipitation for period j	mm
RX3day	Max 3-day precipitation	RX3dayj = max (PRkj), the maximum 3-day precipitation for period j	mm
SDII	No. of rainy days June–September	Total rainfall June–September/No. of rainy days June–September	mm/day

The changing climatic conditions were assessed in this study by using observed and projected daily extreme precipitation data (annual maximum). Trend tests were applied to the observed annual maximum time series to capture any variations. For the 5-, 10-, 20-, 50-, and 100-year return periods, observed and projected 24 h and 72 h annual maximum time series were used to calculate the return level (precipitation depth in mm).

3. Results

3.1. Observed Precipitations

Precipitation data were collected for fifteen stations within the Gandaki Province and adjusted. The stations' elevation ranges from 150 to 3420 m above sea level, and the reported precipitation similarly exhibited a wide range. During the research period, the recorded annual daily maximum precipitation was 357 mm at Pokhara airport stations. Figure 3 shows the trend of observed annual daily extreme precipitation from 2003 to 2020 in the Baglung and Lamjung districts of Gandaki Province.

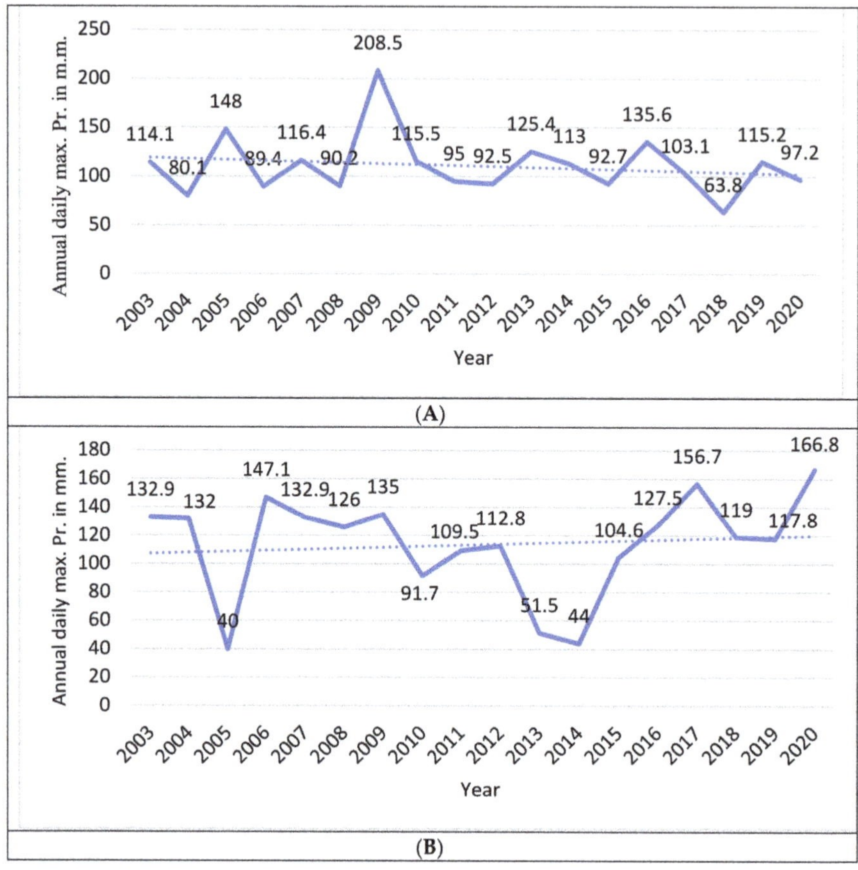

Figure 3. Observed annual daily maximum precipitation trend of (**A**) Baglung and (**B**) Lamjung.

Baglung, Bandipur, Manang Bhot, Tatopani, and Damauli were among fifteen stations that showed a decreasing trend in annual daily extreme precipitation from 2003 to 2020, whereas Kushma, Bobang, Beluwa Girwari, Dumkauli, Khudi Bazar, Gorkha, Jomsom, Syangja, Pokhara Airport, and Lumle demonstrated the increase in precipitation. Lumle station's data were analyzed for 51 years, and for the Pokhara airport station, we analyzed 53 years of data (see Appendix A, Figures A1–A12 and Appendix A).

3.2. Future Climate Projection

This study looked at relevant extreme indices that were "user-relevant" for assessing potential future changes in extreme events. The indices for the extreme-event analysis were calculated over three 20-year periods, using the two GCMs' CMIP6 climate projections. For both GCMs, the first 20 years are the base period, or the control run from 1995 to 1914. The effects of climate change on extreme events were projected for the near future (2021–2040) and the far future (2081–2100). The projections of two GCMs were found to be slightly different from the rest, and it is believed that such discrepancy could be partly due to the raw data resolution, which was relatively coarse compared to the raw data resolution for the base period and projections, but also to the overall extreme precipitation increases over Gandaki Province in comparison to the base period. Future projections from various indices are further discussed.

3.2.1. R50 mm Index

The R50 mm index, also known as really heavy precipitation days, is the number of days in a certain period when the daily precipitation was greater than 50 mm. In comparison to the base period for the SSP2–4.5 scenarios of BCC-CSM2-MR analyzed, the R50 mm was increased by about 9% and 13% in the near future (2021–2040) and far future (2081–2100), respectively (see Figure 4A). On the other hand, from the model MRI-ESM2-0, it is projected to decrease by 4% in the near future and dramatically increase by 122% in the far future (2081–2100) (see Figure 4B).

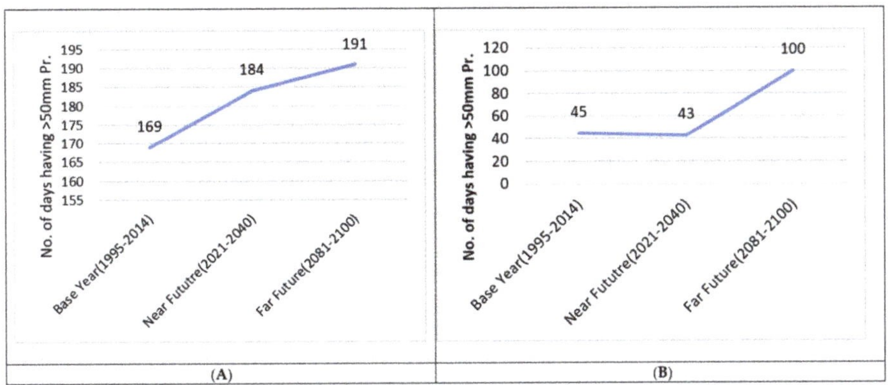

Figure 4. Number of days having greater than 50 mm precipitation (**A**) from the BCC-CSM2-MR model and (**B**) from the MRI-ESM2-0 model.

3.2.2. R100 mm Index

The R100 mm index, also known as very heavy precipitation days, counts the number of days in a given period when the daily precipitation exceeds 100 mm. Figure 5A shows that the R100 mm was increased by around 38.7% and 60% in the near future and far future, respectively, with respect to the base year, from the BCC-CSM2-MR model. According to the MRI-ESM2-0 model, there was no precipitation observed in the base year, while the near future showed 4 days and the far future indicated 13 days (see Figure 5B). We can conclude from this result that both models predicted an increase in very heavy precipitation days in the future, i.e., 2081–2100.

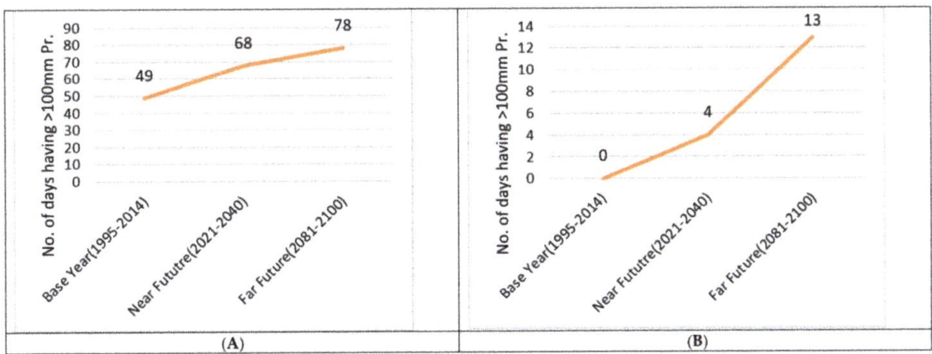

Figure 5. Number of days having greater than 100 mm precipitation (**A**) from the BCC-CSM2-MR model and (**B**) from the MRI-ESM2-0 model.

3.2.3. RX1day Index

The RX1day index represents the maximum daily precipitation within a given period. The return period was utilized to determine the extreme precipitation projection in this index. Equation (1) can be used to calculate the return period by using the graphical analysis:

$$T = (n + 1)/m \tag{1}$$

where n = No. of years, m = rank of precipitation, and T = return period.

We may discover the equation by plotting the return period on the *x*-axis and the yearly daily maximum precipitation in descending order on the *y*-axis in logarithmic form. From this equation, we can estimate future precipitation. Figure 6 shows the sample of the logarithmic equation for the BCC-CSM2-MR model in the base year.

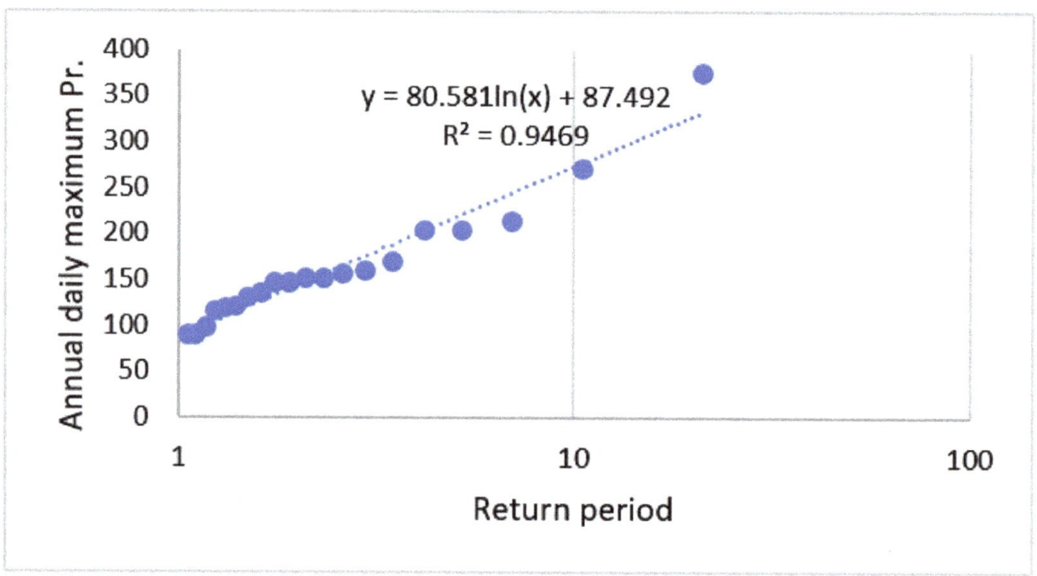

Figure 6. Logarithmic equation for the BCC-CSM2-MR model in the base year.

With reference to the base period, all the stations showed a 11.86%, 11.99%, 12.07%, 12.15%, and 12.19% increase in 5-, 10-, 20-, 50-, and 100-year precipitation, respectively, in the near future. Similarly, a 10.64% and 3.94% increase in the next 5 and 10 years in the far future, respectively, was shown, as well as a 0.49%, 4.47%, and 6.63% decrease in the next 20, 50, and 100 years, from the BCC-CSM2-MR model (see Figure 7A). From the MRI-ESM2-0 model, the results were quite different, in that 5-, 10-, 20-, 50-, and 100-year annual maximum precipitation increased by 28.96%, 41.17%, 50.68%, 60.44%, and 66.31%, respectively, in the near future. Increases in 5, 10, 20, 50, and 100 years in the far future are 50.54%, 52.97%, 54.87%, 56.81%, and 57.99%, which are slightly less than the near-future precipitation for the 100-year return period (see Figure 7B).

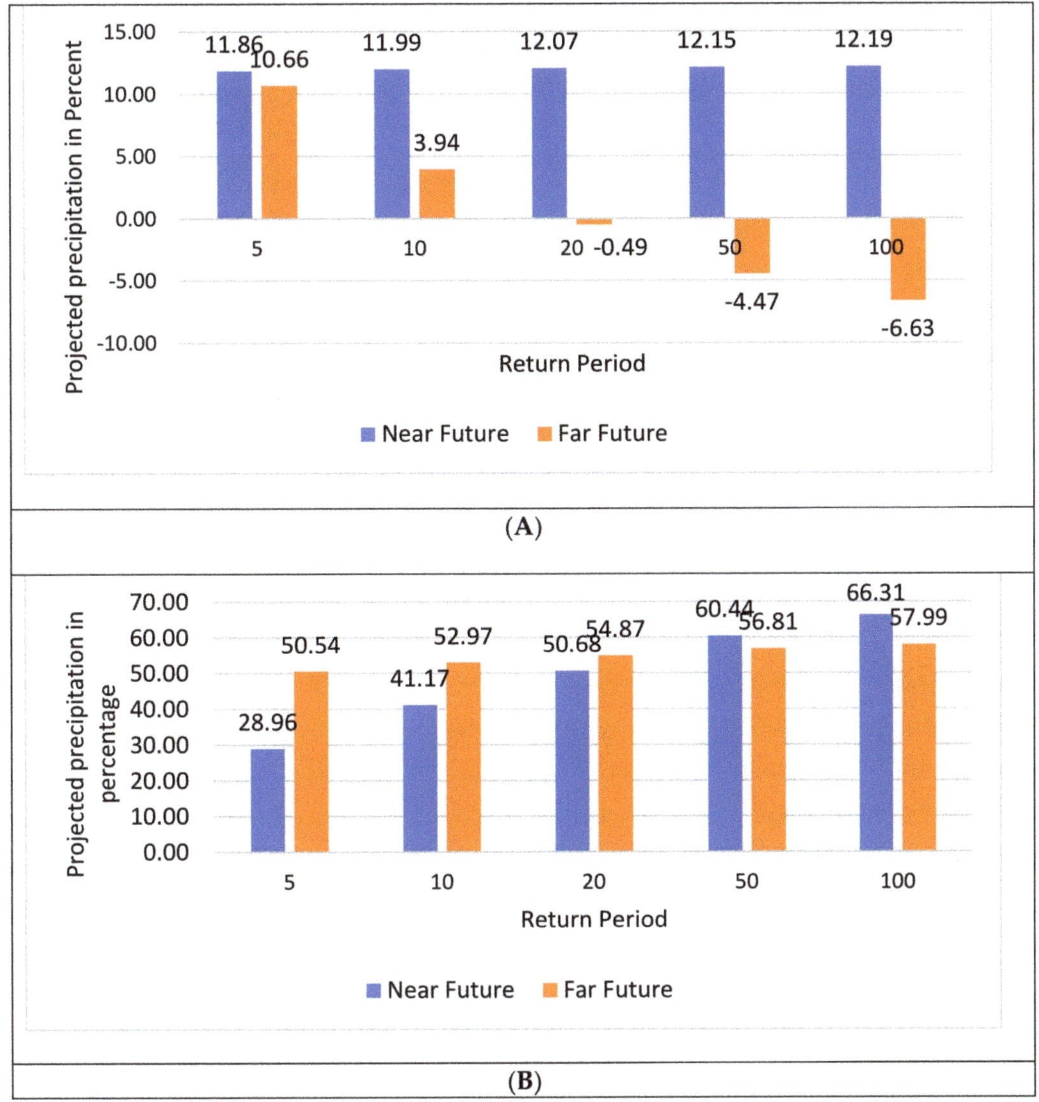

Figure 7. The projected daily extreme precipitation percentage (**A**) from the BCC-CSM2-MR model and (**B**) from the MRI-ESM2-0 model.

Figure 8 shows the spatial distribution of annual daily extreme precipitation for the 100-year return period output from the graphical analysis from the MRI-ESM2-0 model, under the SSP2–4.5 and SSP5–8.5 scenarios for the near future (2021–2040) relative to 1995–2014. ArcGIS was used to visualize the observed and projected extreme precipitation with a cell size of 100 m over the Gandaki Province. As shown in the figure, the blue color indicates the highest precipitation among the study areas, whereas the red color shows the lowest precipitation.

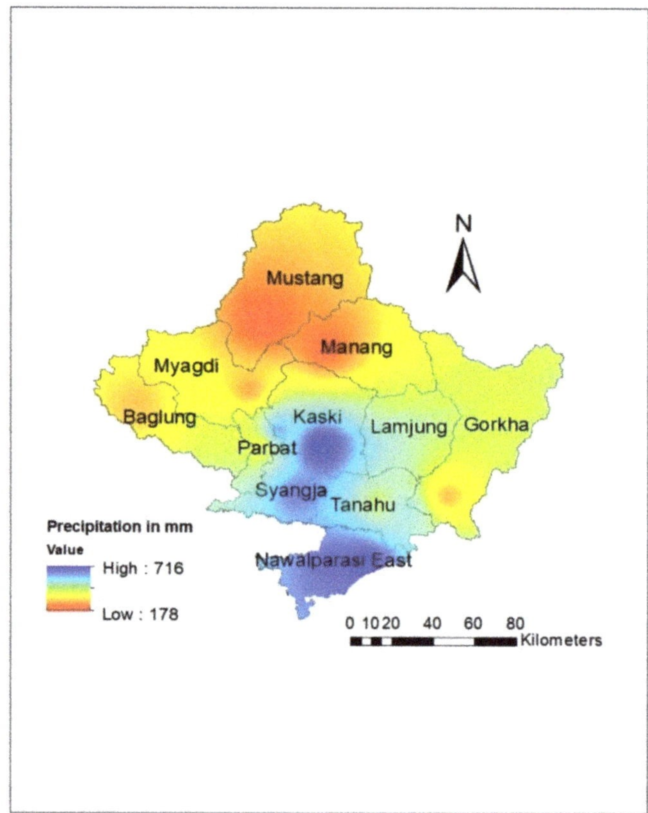

Figure 8. The 100-year return period's precipitation from the MRI-ESM2-0 model for the near future.

Extreme precipitation for 5-, 10-, 20-, 50-, and 100-year return periods was observed to be higher in Nawalparasi East, Kaski, Syangja, and the Western parts of the Tanahun district, respectively, whereas the Mustang and Manang districts showed the least precipitation. Similarly, the BCC-CSM2-MR and MRI-ESM2-0 models show higher precipitation in Nawalparasi East, Kaski, Syangja, and the western part of Tanahun. The Mustang and Manang districts showed the minimum precipitation (see Appendix B, Figure A15).

3.2.4. RX3day Index

The RX3day index is similar to the RX1day index, but it measures the most precipitation observed in three consecutive days during a specific time period, making it another precipitation index that is appropriate for identifying flood-causing storm events. Figure 9 shows that RX3day was projected to increase by about 12.31%, 11.81%, 11.47%, 11.16%, and 10.99% in the near future (2021–2040) and by 16.28%, 13.35%, 11.35%, 9.53%, and 8.52% in the far future (2081–2100) for 5, 10, 20, 50, and 100 years, respectively, from the BCC-CSM2-MR model. Meanwhile, MRI-ESM2-0 showed a 22.16%, 36.33%, 47.73%, 59.78%, and 67.21% increase in the near future (2021–2040), with respect to the base-year, (1995–2014) for 5-, 10-, 20-, 50-, and 100-year return periods and 56.97%, 68.36%, 77.52%, 87.20%, and 93.17% in the far future (2081–2100), respectively.

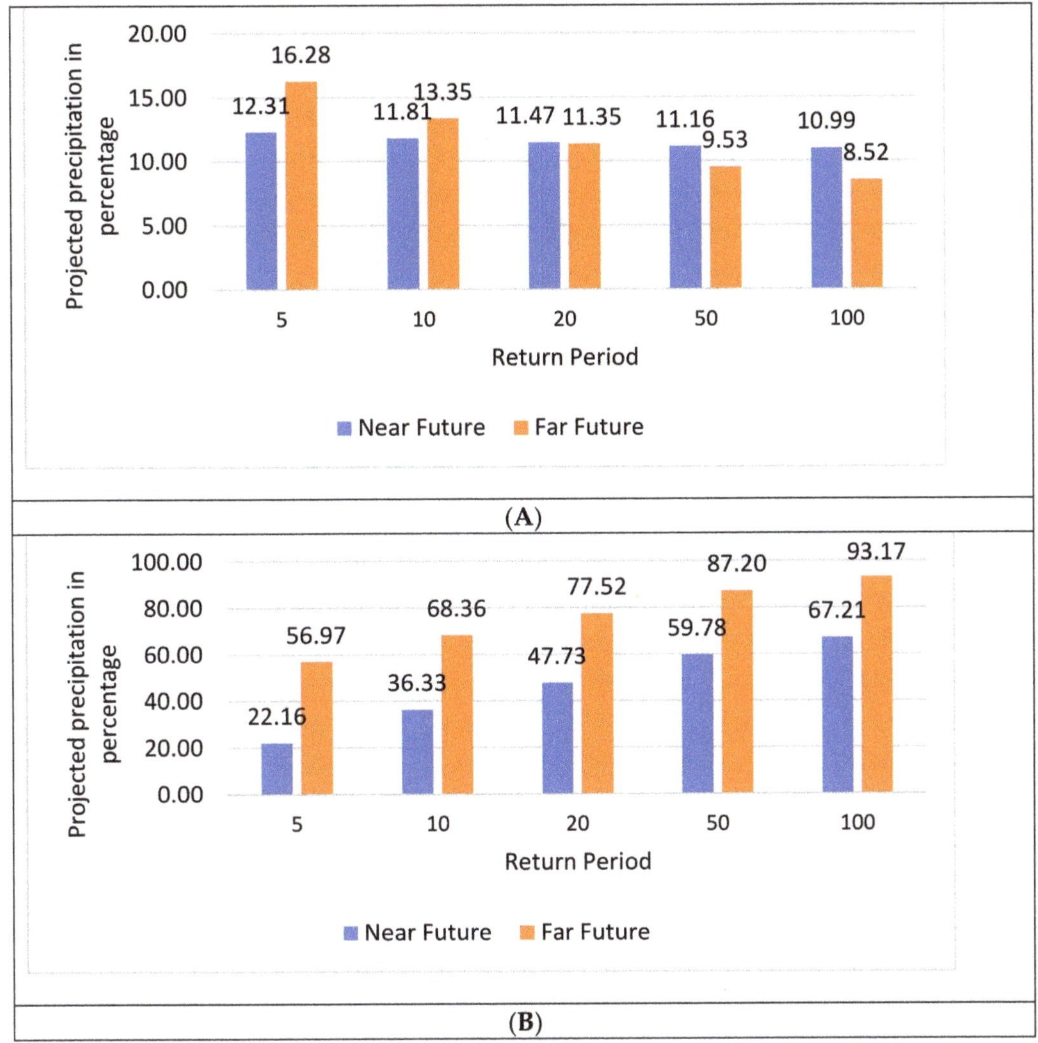

Figure 9. The projected 3-day extreme precipitation percentage (**A**) from the BCC-CSM2-MR model and (**B**) from the MRI-ESM2-0 model.

Similarly, Figure 10 shows the spatial distribution of annual three-day extreme precipitation for a 100-year return-period output from the graphical analysis from the MRI-ESM2-0, under the SSP2–4.5 and SSP5–8.5 scenarios for the far future (2081–2100), relative to 1995–2014. ArcGIS was used to visualize the observed and projected extreme precipitation with a cell size of 100 m over the Gandaki Province. As shown in Figure 10, the blue color indicates the highest precipitation among the study areas, whereas the red color shows the lowest precipitation.

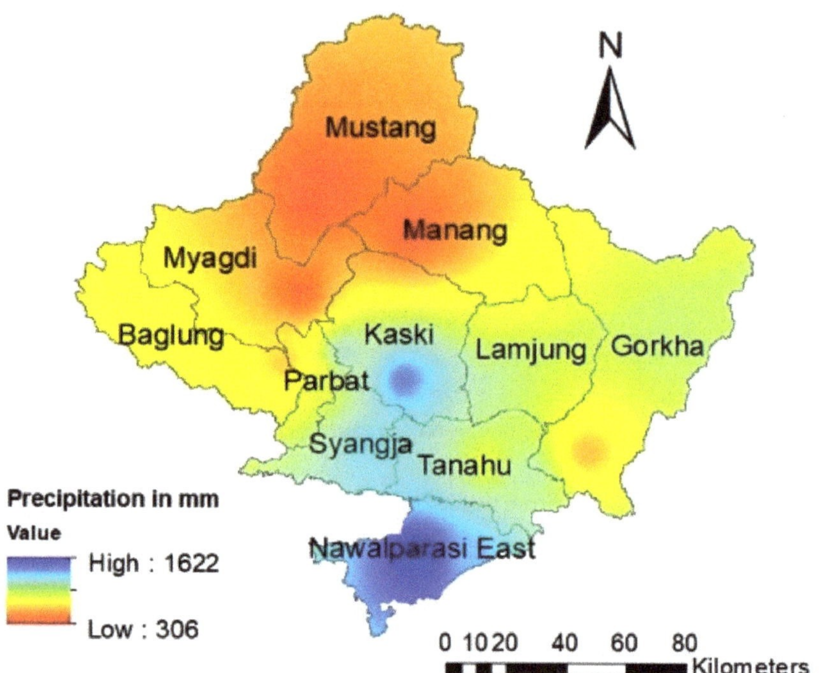

Figure 10. Annual 3-day extreme precipitation for 100-year return periods from the MRI-ESM2-0 model for the far future.

Extreme precipitation for 5-, 10-, 20-, 50-, and 100-year return periods was predicted to be higher in Nawalparasi East, Kaski, Syangja, and the Western parts of the Tanahun district, respectively, whereas the Mustang and Manang districts showed the least precipitation. The BCC-CSM2-MR and MRI-ESM2-0 models showed higher precipitation in Nawalparasi East, Kaski, Syangja, and the western part of Tanahun. The Mustang and Manang districts showed the lowest precipitation (see Appendix B, Figure A16). The highest precipitation for annual three-day extreme precipitation was projected to be 1622 mm.

3.2.5. SDI Index

The simple precipitation intensity index was computed by taking the sum of precipitation on wet days (days with >1 mm of precipitation) and dividing that by the number of wet days in the period. This gives the mean precipitation on wet days. We computed wet days only from June to September in this study because June–September is the rainy season in this province.

The SDII in the base year was 15 mm/day, which is considered to increases by about 3% and 17% in the near future (2021–2040) and far future (2081–2100), respectively, based on the BCC-CSM2-MR model shown in Figure 11A. The SDII, on the other hand, was projected to increase by 2% and 44% in the near future (2021–2040) and far future (2081–2100), respectively, based on MRI-ESM2-0, where the SDII in the base year is 10.33 mm/day as shown in Figure 11B.

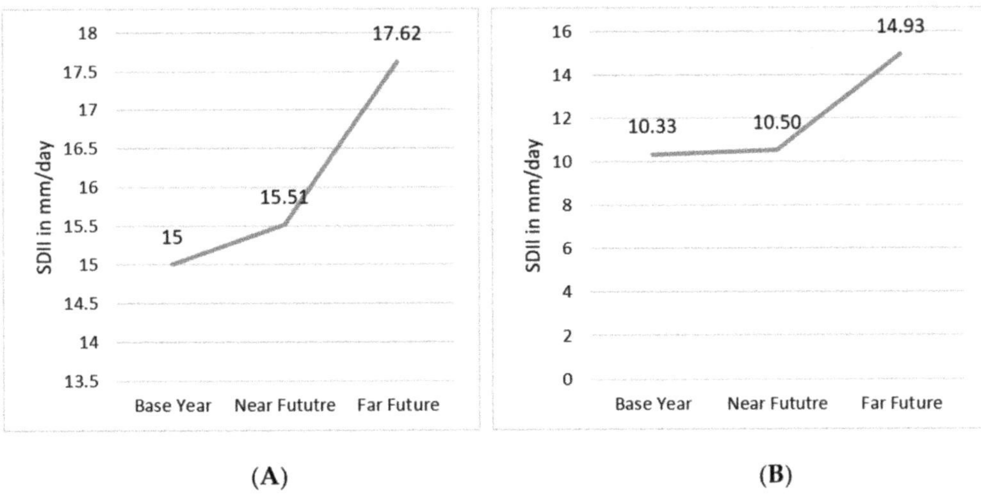

(A)   **(B)**

Figure 11. SDII for June–September (**A**) from BCC-CSM2-MR model and (**B**) from MRI-ESM2-0 model.

4. Discussion

Using observed data and GCMs data produced from CMIP6, this study explored projected changes in precipitation extremes across the Gandaki Province, Nepal. The study used a subset of ETCCDI's precipitation indices to examine extreme events in two future timelines: the near future (2021–2040) and the far future (2081–2100). From the various studies, CMIP6, MIP, SSP2–4.5, and SSP5–8.5, scenarios were used for the projection of extreme events over the study area [16]. A Multi-Model Ensemble (MME) enables the clear identification of signal of interest; hence, it spreads the natural variability. Many studies praised CMIP6-MME's ability to accurately reproduce precipitation extremes when compared to the use of individual models in diverse domains [17].

An examination of extreme estimates revealed the region will have more wet days during the near future (2021–2040) and far future (2081–2100) relative to the baseline period in June–September. An increase of R100 mm around the study area for all regions. It indicated an intensification of very wet events, such as flashfloods, floods, and landslides over the region. Both the increase in RX1day and RX3day precipitation could pose a critical threat to the socioeconomic condition of the region and the government for water-related disaster management in the future. In other words, future extreme precipitation is expected to occur more frequently, meaning that the risk of flooding in these river basins is expected to increase in the future. Extreme rainfall events have serious consequences for economic sectors, such as agriculture, and food security, as these are closely linked to climate change. Maximum precipitation was forecasted in Kaski, Nawalparasi East, Syangja, and the western half of the Tanahun region, raising the risk of future floods and water-related calamities. Due to the lack of precipitation, there is a potential for drought in the future for Mustang, Manang, and the upper portion of the Myagdi.

The findings of these analyses could provide an opportunity to better understand future extreme events predicted by GCMs. Several limitations were identified which have an impact on the current results. Such limitations include the use of few models during the analysis, the use of a limited number of statistical features, the use of a limited number of indices, and the lack of examination of physical mechanisms to aid in a better understanding of the phenomenon of extreme occurrence. GCM precipitation outputs are reported with greater uncertainties, such as too many wet days and smaller intensities. However, a clear characterization of the potential extremes defining the intensity, frequency, and duration remains critical for various stakeholders to make better-informed decisions. More research is needed to investigate the underlying physical features that influence the occurrence of extreme incidences projected for relevant policies, as well as the use of bias-corrected models to reduce inter-model biases. In this study, GCM precipitation output patterns of near-future and far-future periods with respect to the base period were analyzed to characterize climate-change impacts on precipitation output. Statistical and dynamic downscaling techniques are popularly carried out to minimize uncertainties and enhance the quality of climate outputs. The bias-correction technique, a statistical downscaling technique, is applied to minimize discrepancies in the wet days and intensities. These techniques need to be explored and applied for reliable estimates for climate-change impacts on extreme precipitation characteristics.

These findings suggest that water-related disasters could be common in the study area, and that they are likely to become more common in the future, as a result of pronounced climate change, which will either increase precipitation or change the nature of precipitation so that it is short-lived and high in intensity. It is better to prepare for mitigation and adaptation measures to deal with the adverse situation in the future by reducing the disaster.

5. Conclusions

The study investigated the potential impact of climate change on future extreme precipitation over the Gandaki Province, using observed and the latest GCMs of CMIP6 with 100 km resolution. The analysis was carried out by using five extreme precipitation indices defined by ETCCDI. They are R50 mm, R100 mm, SDII, RX1day, and RX3day.

For the JJAS season, with respect to the base year (1995–2014), SDII indices were projected to increase in both the near future (2021–2040) and the far future (2081–2100) from the BCC-CSM2-MR model, whereas the MRI-ESM2-0 model shows a dramatic increase in the far future (2081–2100). However, RX50 mm was projected to be slightly lower in the far future (2081–2100), as compared to the near future (2021–2040), with MRI-ESM2-0. There was no RX100 mm observed in the base year, but there was an increase in both the near future (2021–2040) and the far future (2081–2100). Similarly, both the RX1day and RX3day showed an extensive increase in precipitation in the near future and far future. Overall, the indices indicated that future extreme precipitation in the study area would increase with respect to the base year (1995–2014).

The monsoonal circulation of South Asia has a significant impact on Nepal, which is located on the southern edge of the Himalayas [18]. This study showed the more extensive precipitation in the future as compared to a previous study that the downscaling of GCM [11]. On the basis of this study, immediate mitigation and adaptation measures shall be implemented to avoid uncontrollable consequences. Furthermore, further research should be carried out to evaluate the potential population exposure to forecasted precipitation extremes.

Extreme precipitation is increasing in the future, posing floods and landslides, so it is suggested for policymakers and water-related structure designers to consider revised return-period rainfall during the structure design of constructions such as bridges and dams. It is also advised to revise the design guidelines and adopt the suitable mitigation measures.

6. Dedication

This is dedicated to my father, Hum Nath Pandey, who died in the flood.

Author Contributions: Writing–original draft, S.P.; Writing–review & editing, B.K.M. All authors have read and agreed to the published version of the manuscript.

Funding: This research received no external funding.

Institutional Review Board Statement: Not applicable.

Informed Consent Statement: Not applicable.

Data Availability Statement: Not applicable.

Conflicts of Interest: The authors declare no conflict of interest.

Appendix A. Observed Annual Daily Maximum Precipitation of Study Area

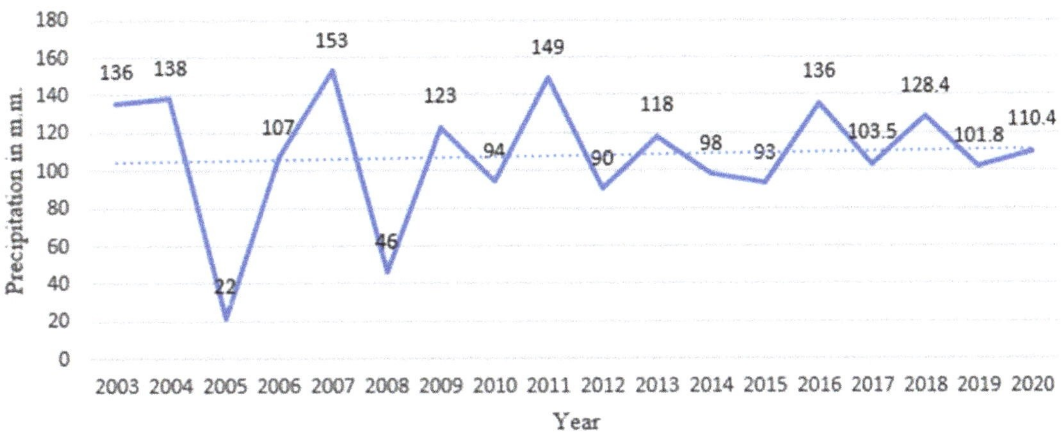

Figure A1. Observed annual daily maximum precipitation of Kushma, Parbat.

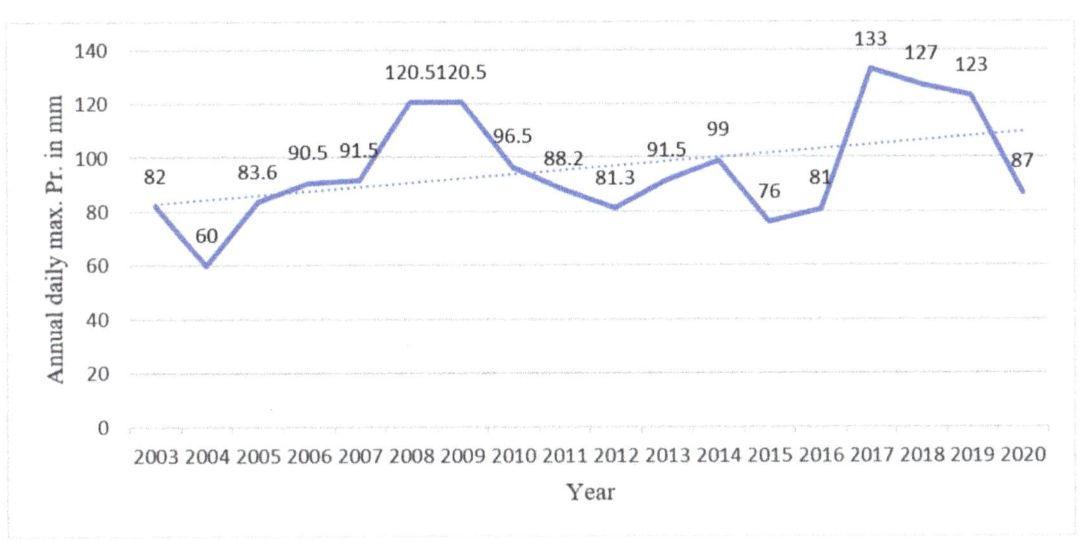

Figure A2. Observed annual daily maximum precipitation of Bobang, Baglung.

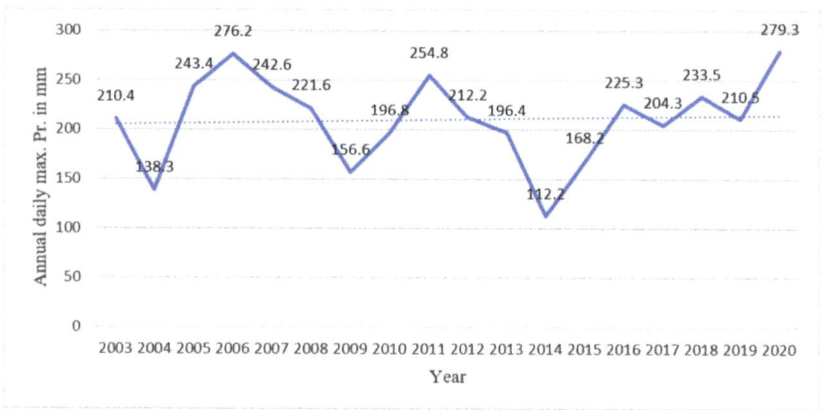

Figure A3. Observed annual daily maximum precipitation of Beluwa Girwari, Nawalpur.

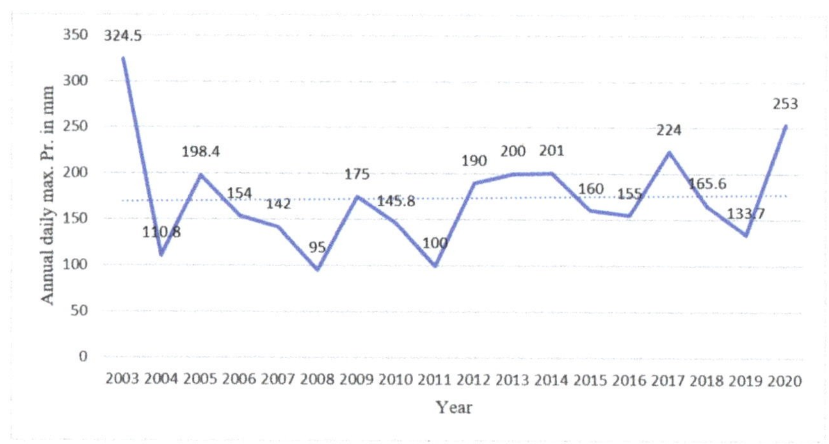

Figure A4. Observed annual daily maximum precipitation of Dumkauli, Nawalpur.

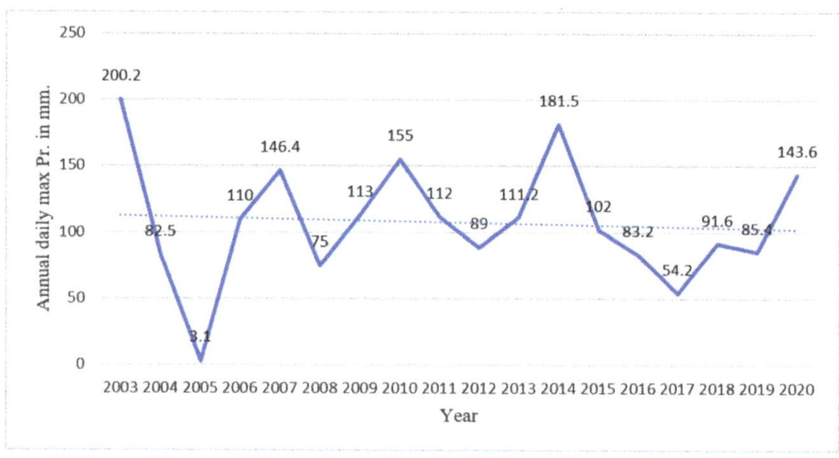

Figure A5. Observed annual daily maximum precipitation of Bandipur, Tanahun.

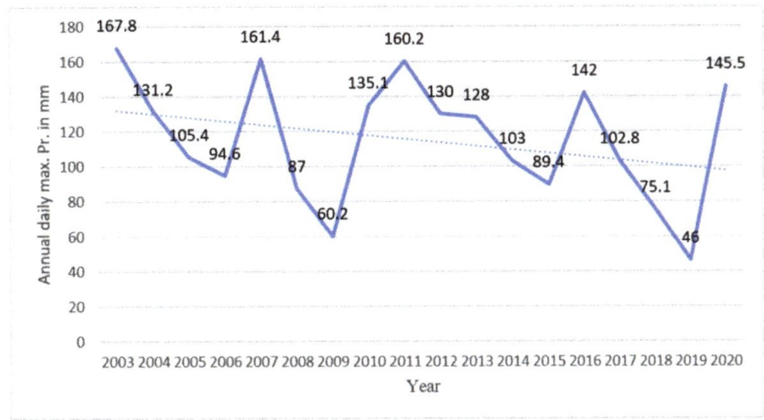

Figure A6. Observed annual daily maximum precipitation of Damauli, Tanahun.

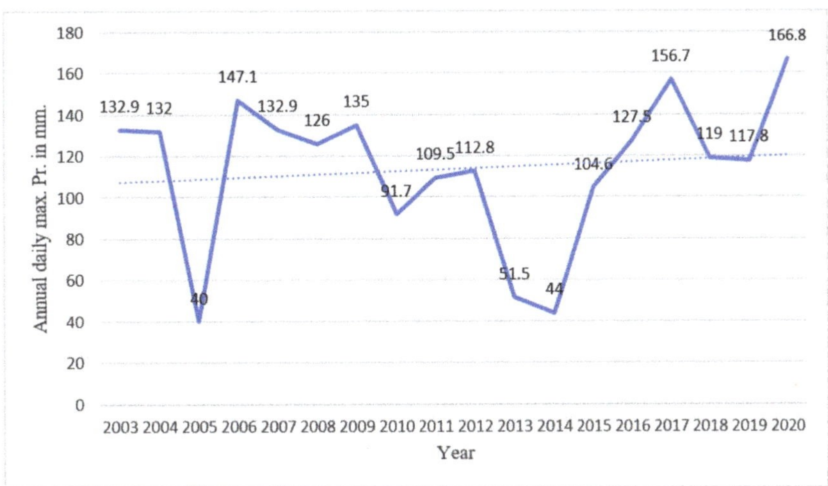

Figure A7. Observed annual daily maximum precipitation of Khudi Bazar, Lamjung.

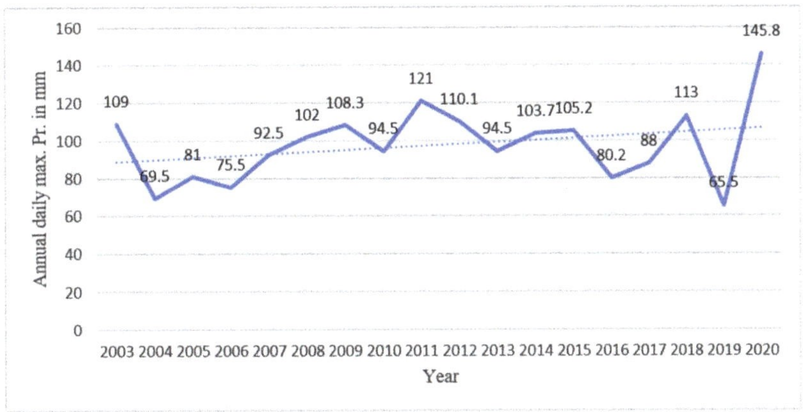

Figure A8. Observed annual daily maximum precipitation of Gorkha.

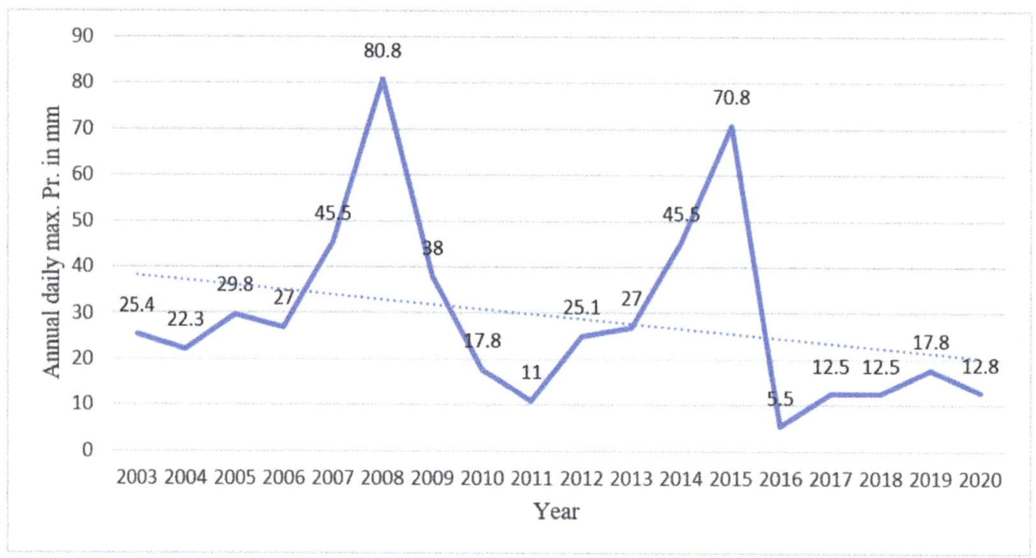

Figure A9. Observed annual daily maximum precipitation of Manang Bhot, Manang.

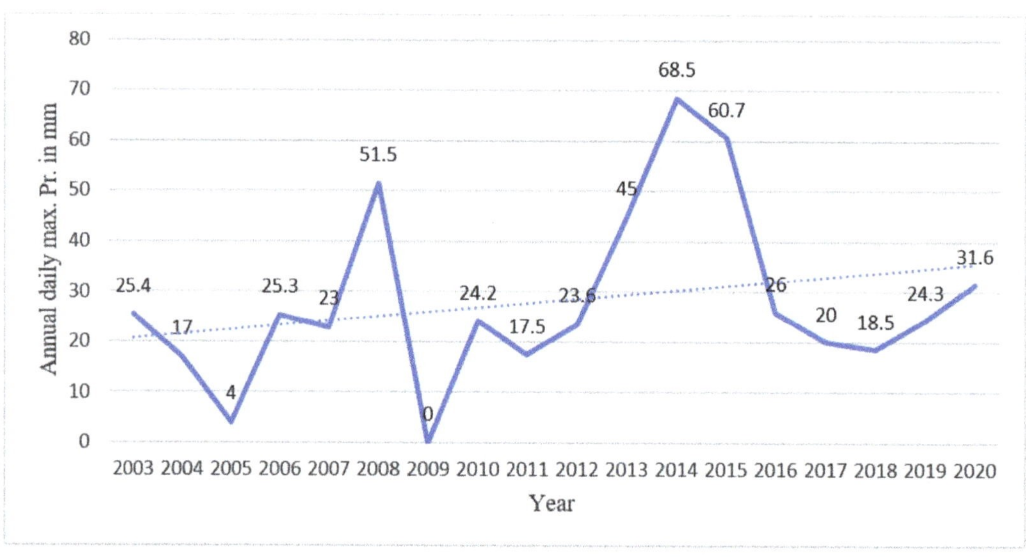

Figure A10. Observed annual daily maximum precipitation of Jomsom, Mustang.

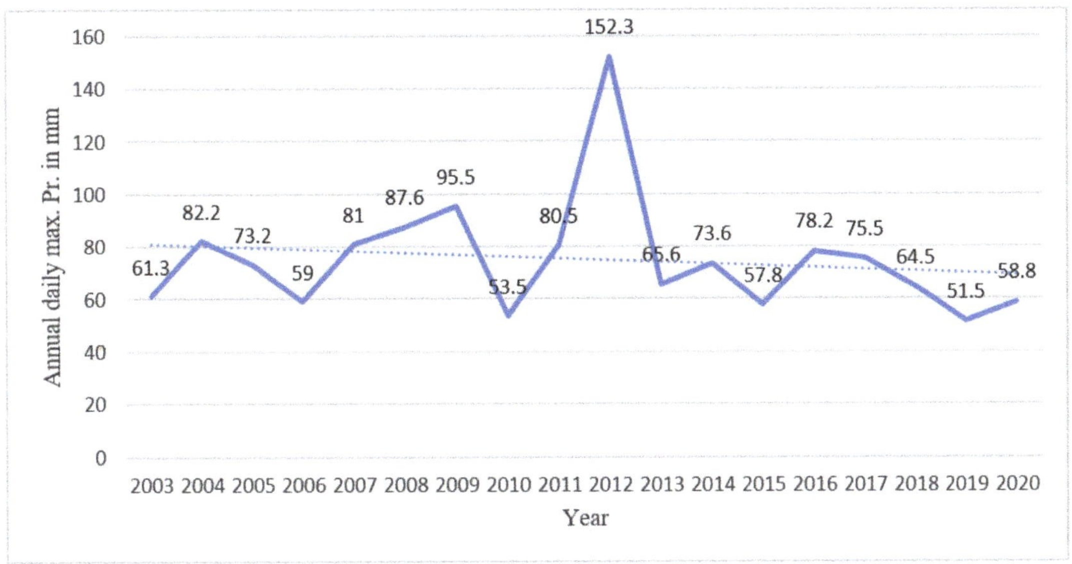

Figure A11. Observed annual daily maximum precipitation of Tatopani, Myadi.

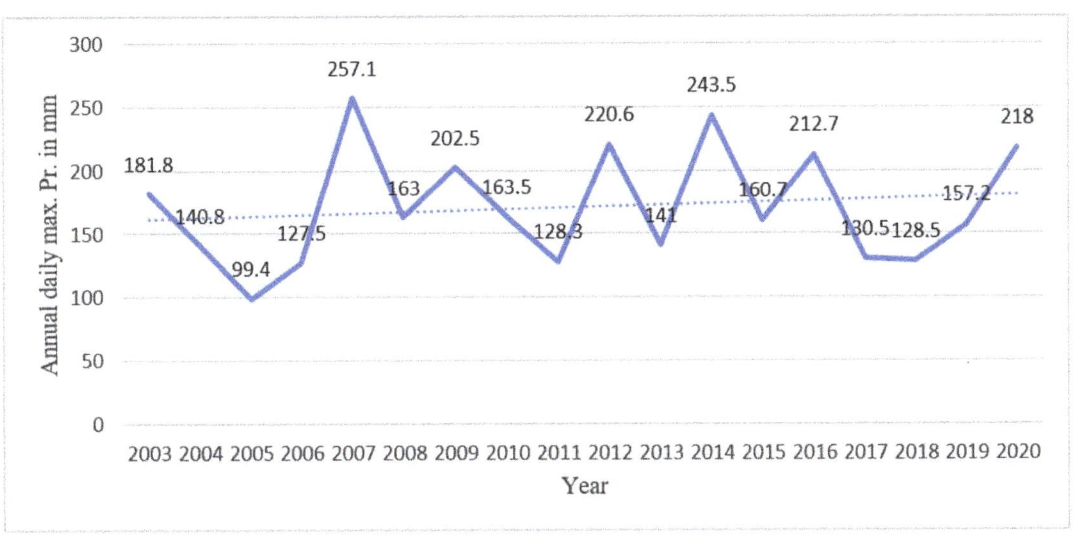

Figure A12. Observed annual daily maximum precipitation of Syangja.

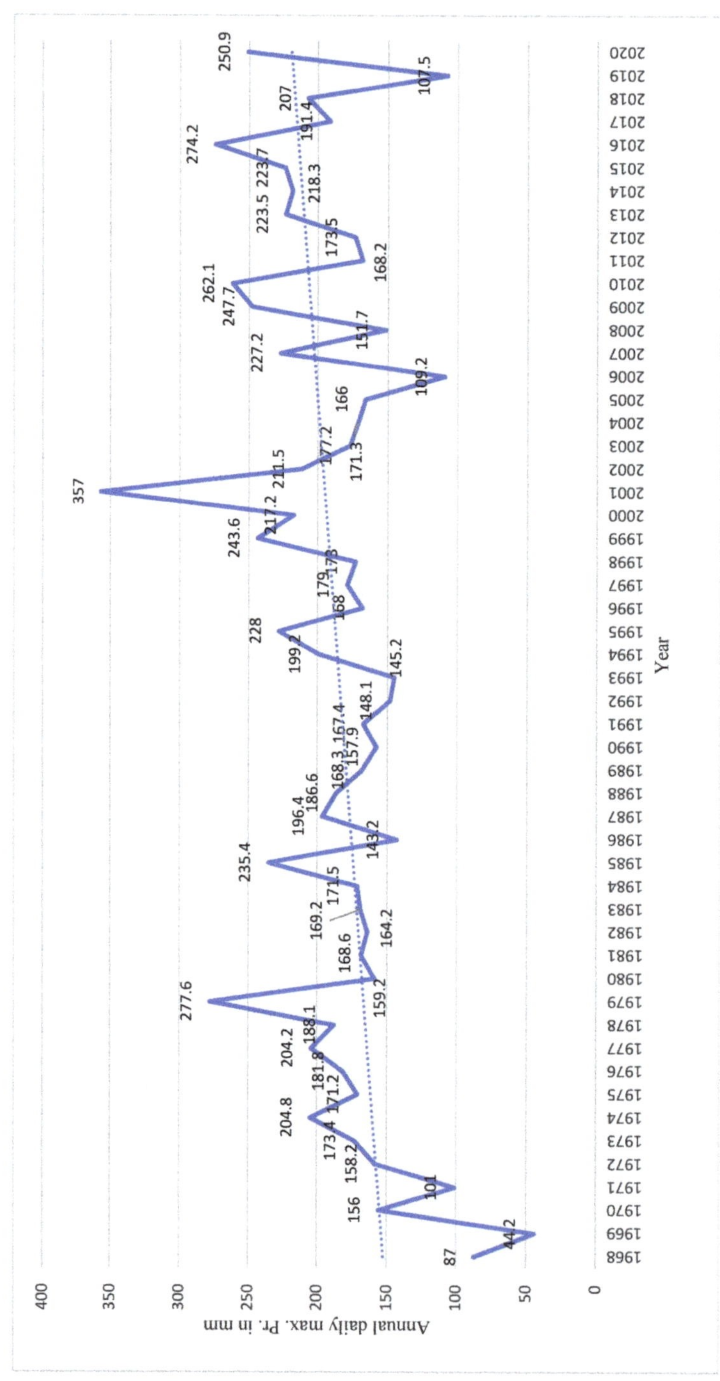

Figure A13. Observed annual daily maximum precipitation of Pokhara Airport, Kaski.

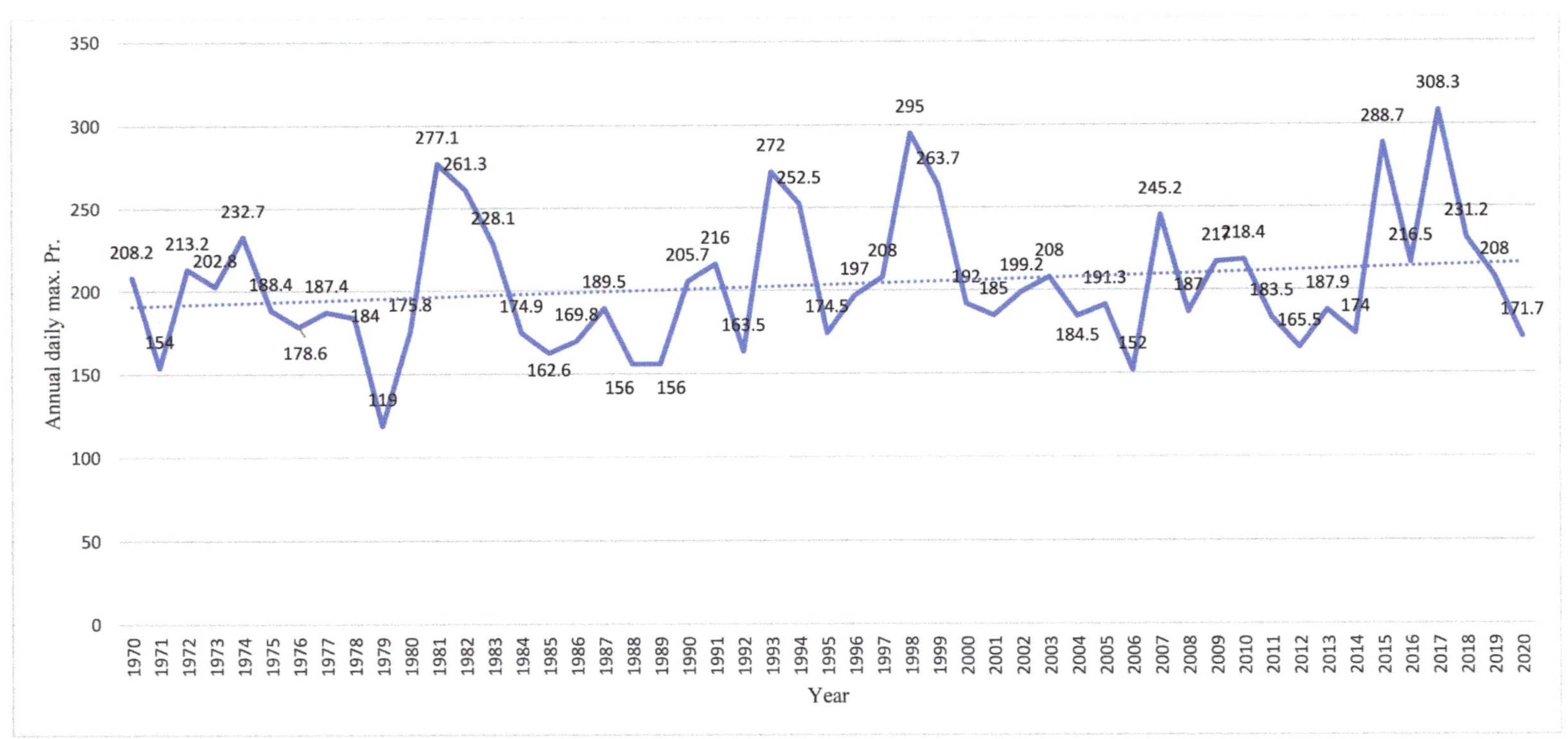

Figure A14. Observed annual daily maximum precipitation of Lumle, Kaski.

Appendix B. Spatial Variation of Forecasted Return Periods Precipitation

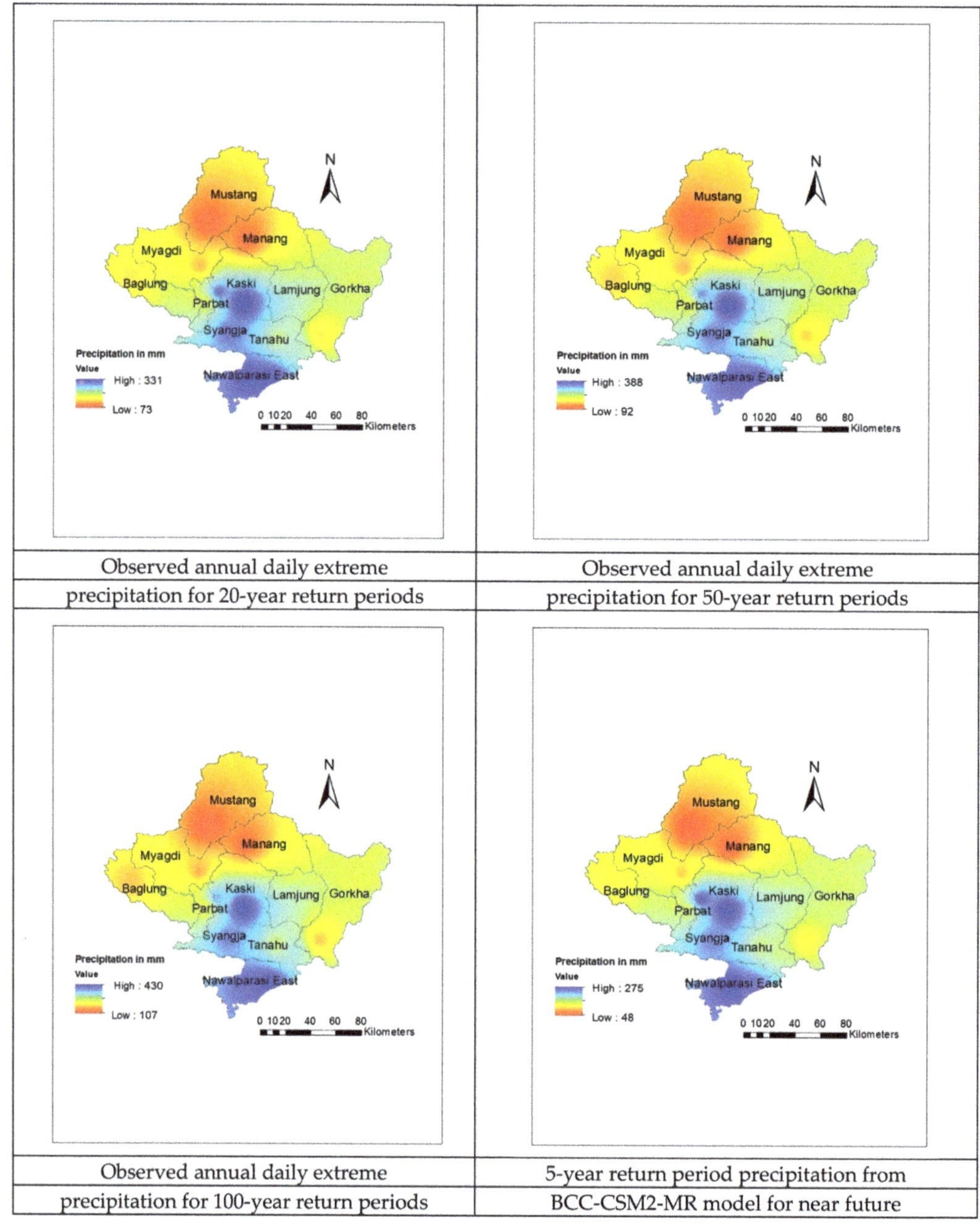

Observed annual daily extreme precipitation for 20-year return periods	Observed annual daily extreme precipitation for 50-year return periods
Observed annual daily extreme precipitation for 100-year return periods	5-year return period precipitation from BCC-CSM2-MR model for near future

Figure A15. *Cont.*

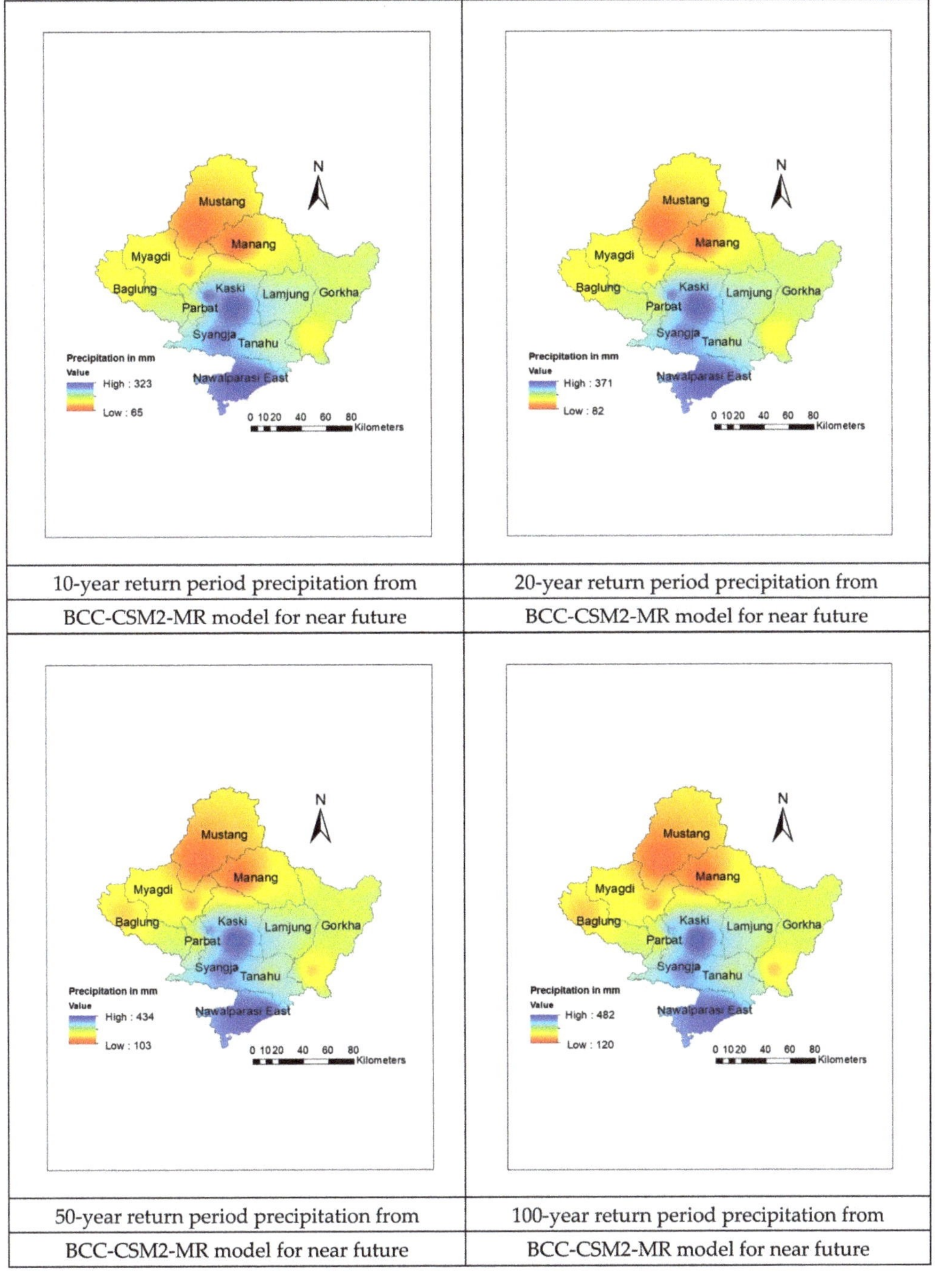

10-year return period precipitation from BCC-CSM2-MR model for near future	20-year return period precipitation from BCC-CSM2-MR model for near future
50-year return period precipitation from BCC-CSM2-MR model for near future	100-year return period precipitation from BCC-CSM2-MR model for near future

Figure A15. *Cont.*

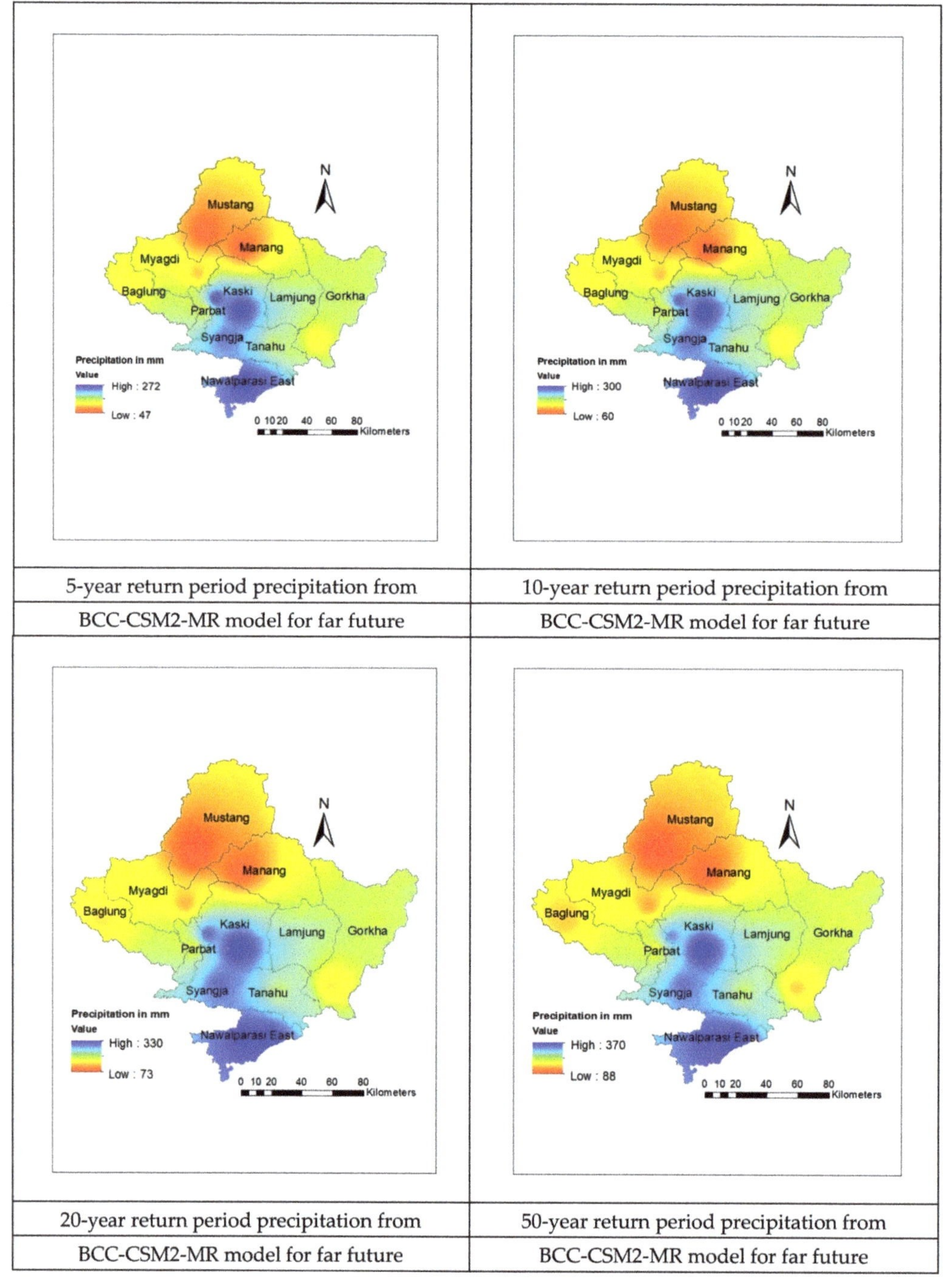

5-year return period precipitation from BCC-CSM2-MR model for far future	10-year return period precipitation from BCC-CSM2-MR model for far future
20-year return period precipitation from BCC-CSM2-MR model for far future	50-year return period precipitation from BCC-CSM2-MR model for far future

Figure A15. *Cont.*

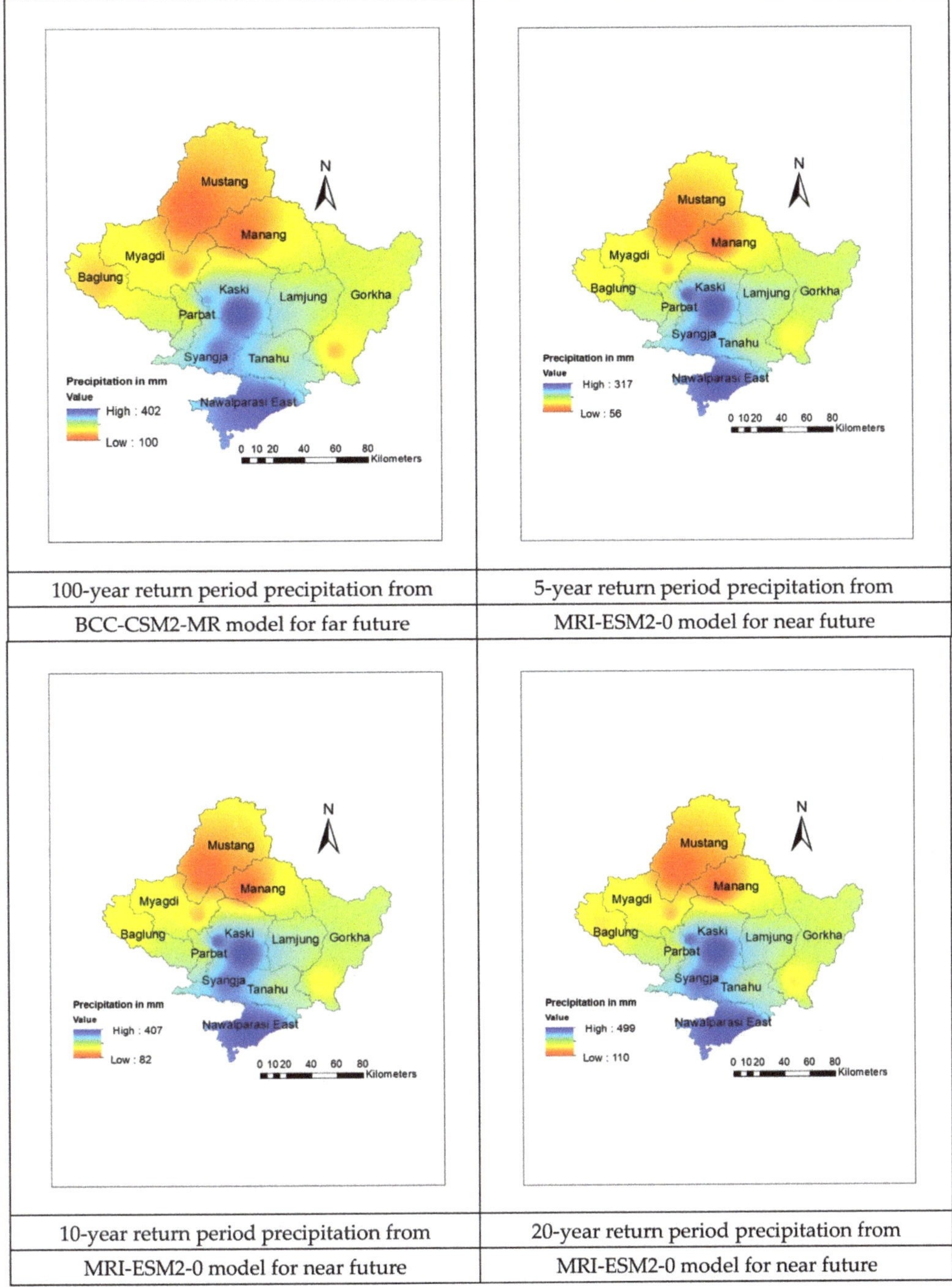

Figure A15. *Cont.*

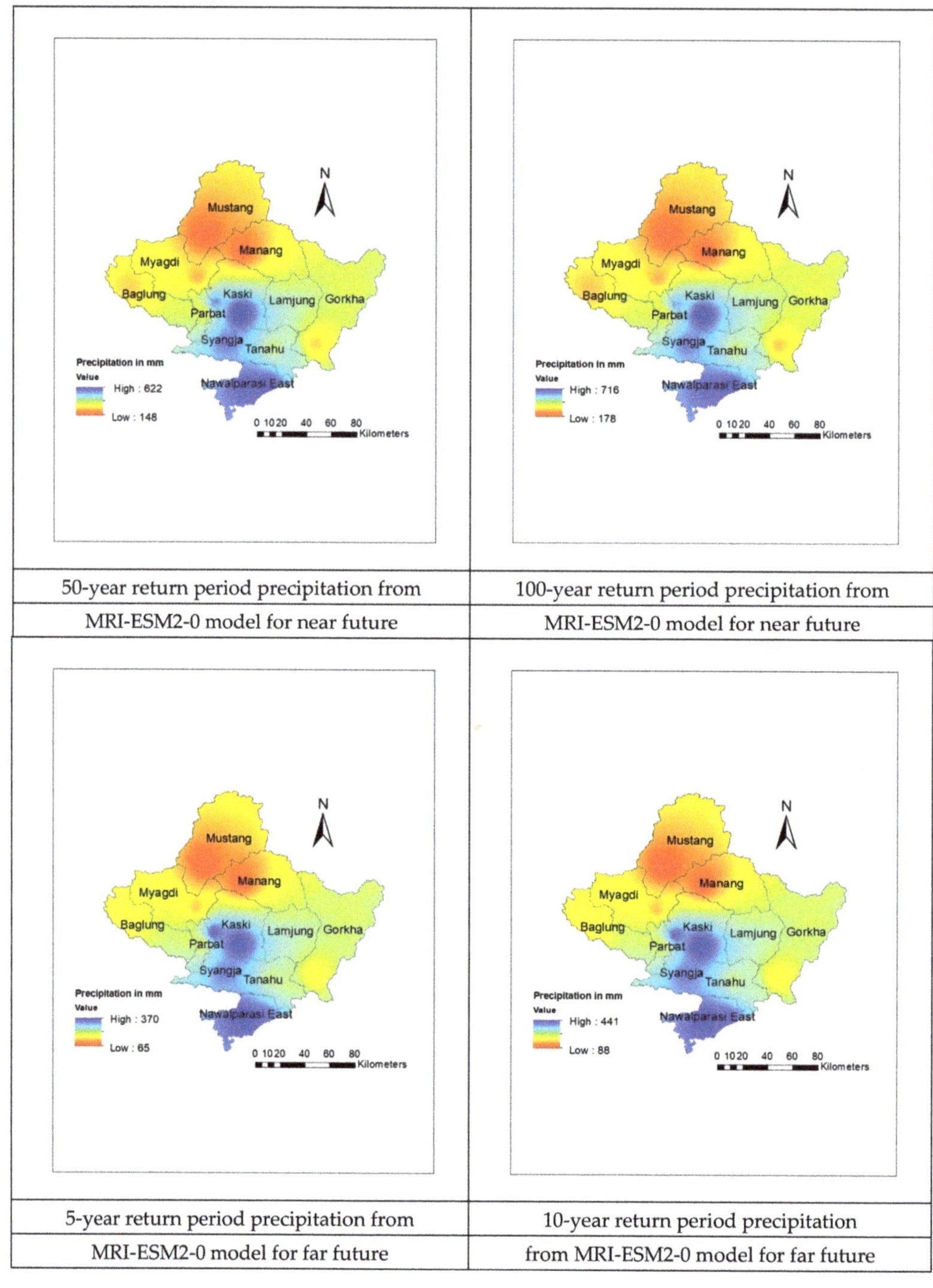

50-year return period precipitation from MRI-ESM2-0 model for near future	100-year return period precipitation from MRI-ESM2-0 model for near future
5-year return period precipitation from MRI-ESM2-0 model for far future	10-year return period precipitation from MRI-ESM2-0 model for far future

Figure A15. *Cont.*

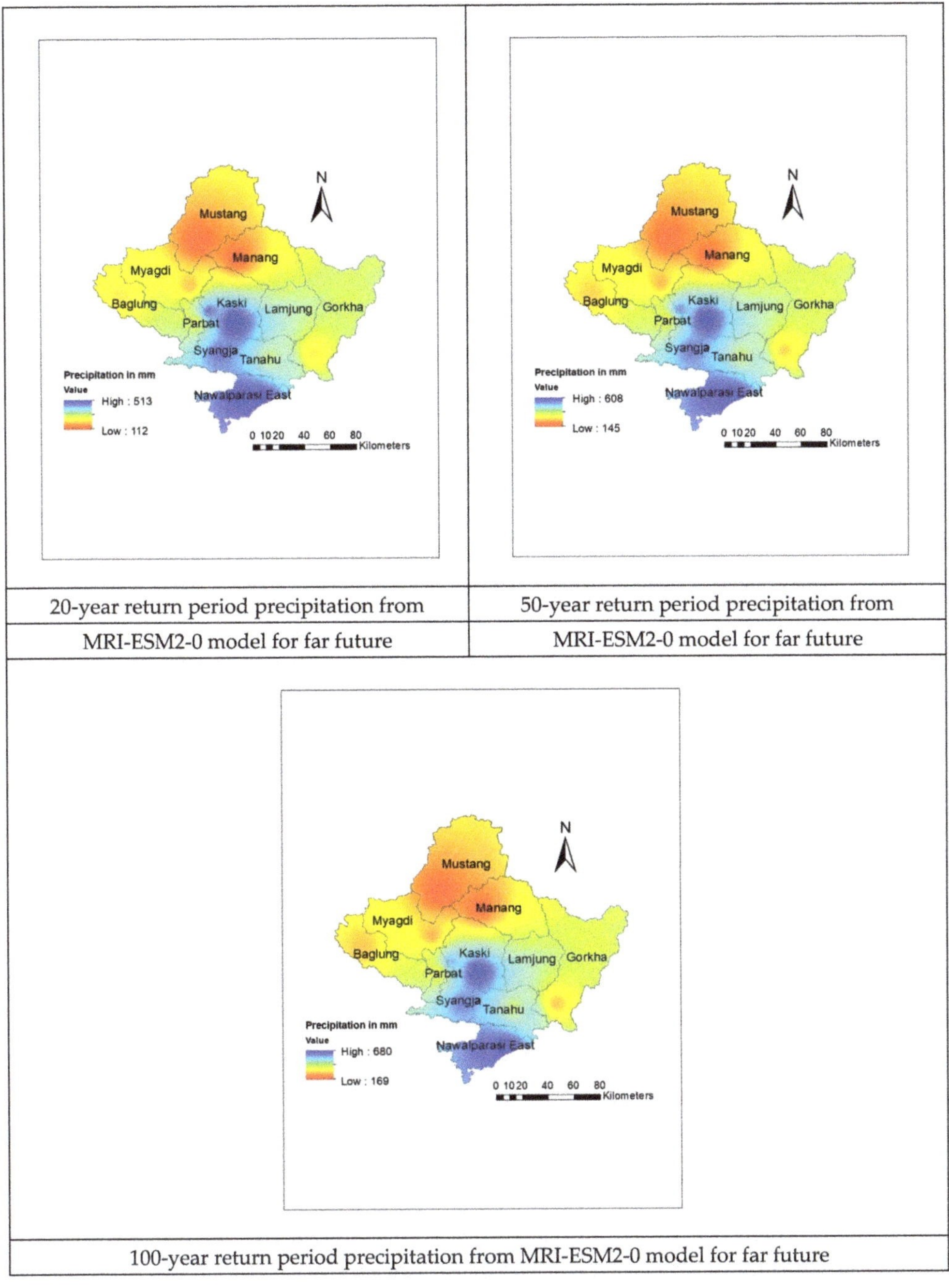

Figure A15. Projected spatial annual daily extreme precipitation.

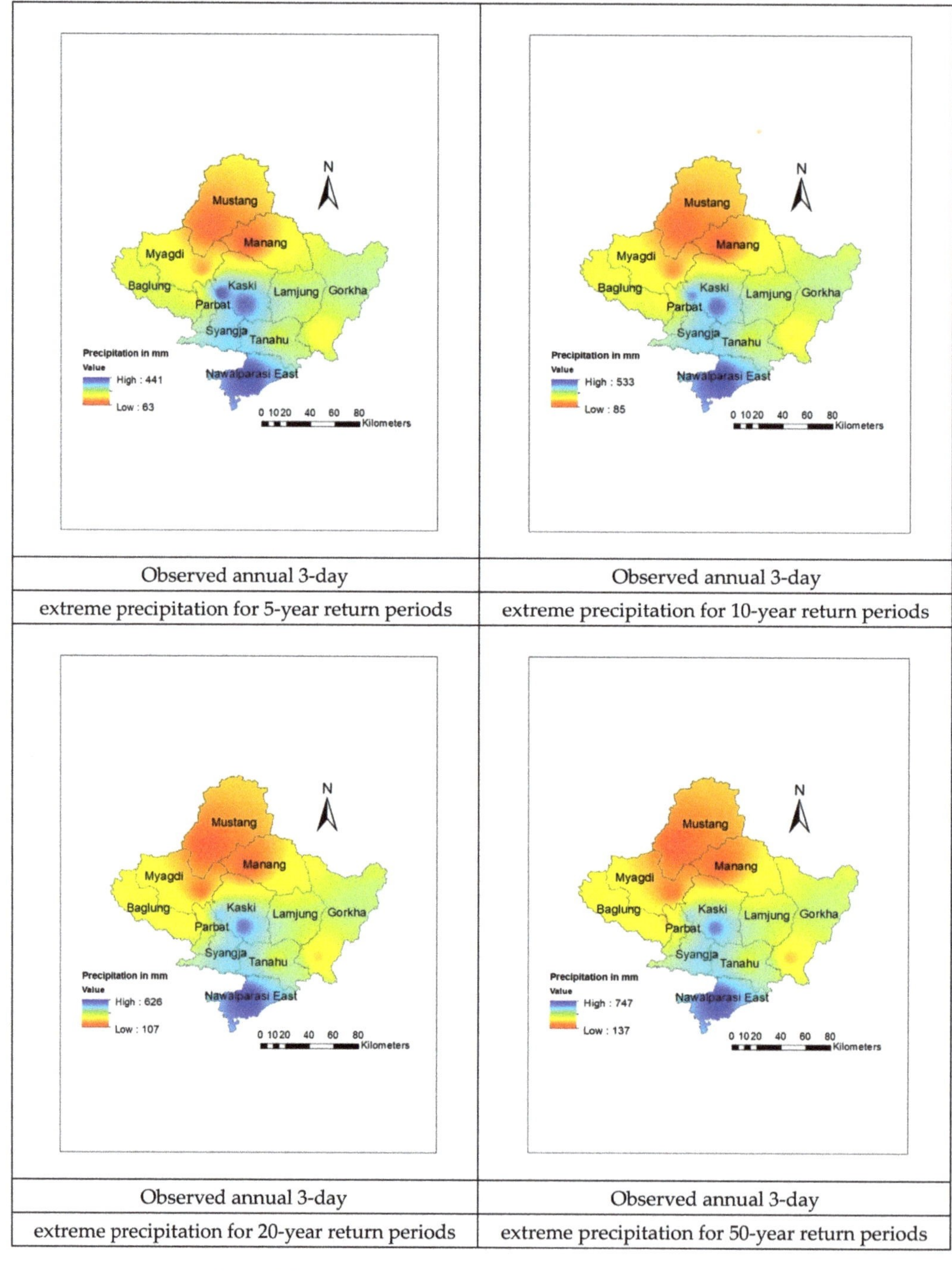

Observed annual 3-day extreme precipitation for 5-year return periods	Observed annual 3-day extreme precipitation for 10-year return periods
Observed annual 3-day extreme precipitation for 20-year return periods	Observed annual 3-day extreme precipitation for 50-year return periods

Figure A16. *Cont.*

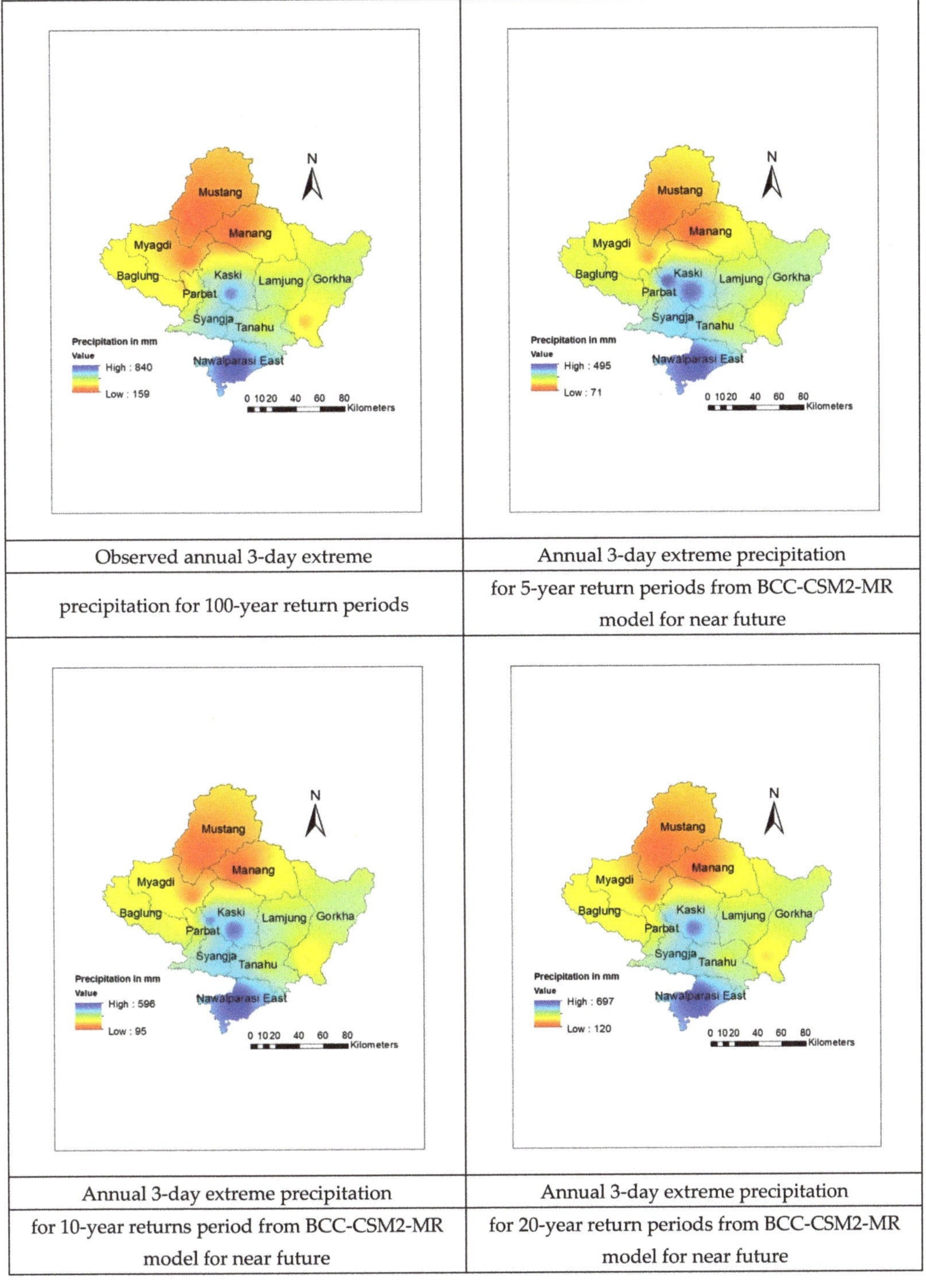

Figure A16. *Cont.*

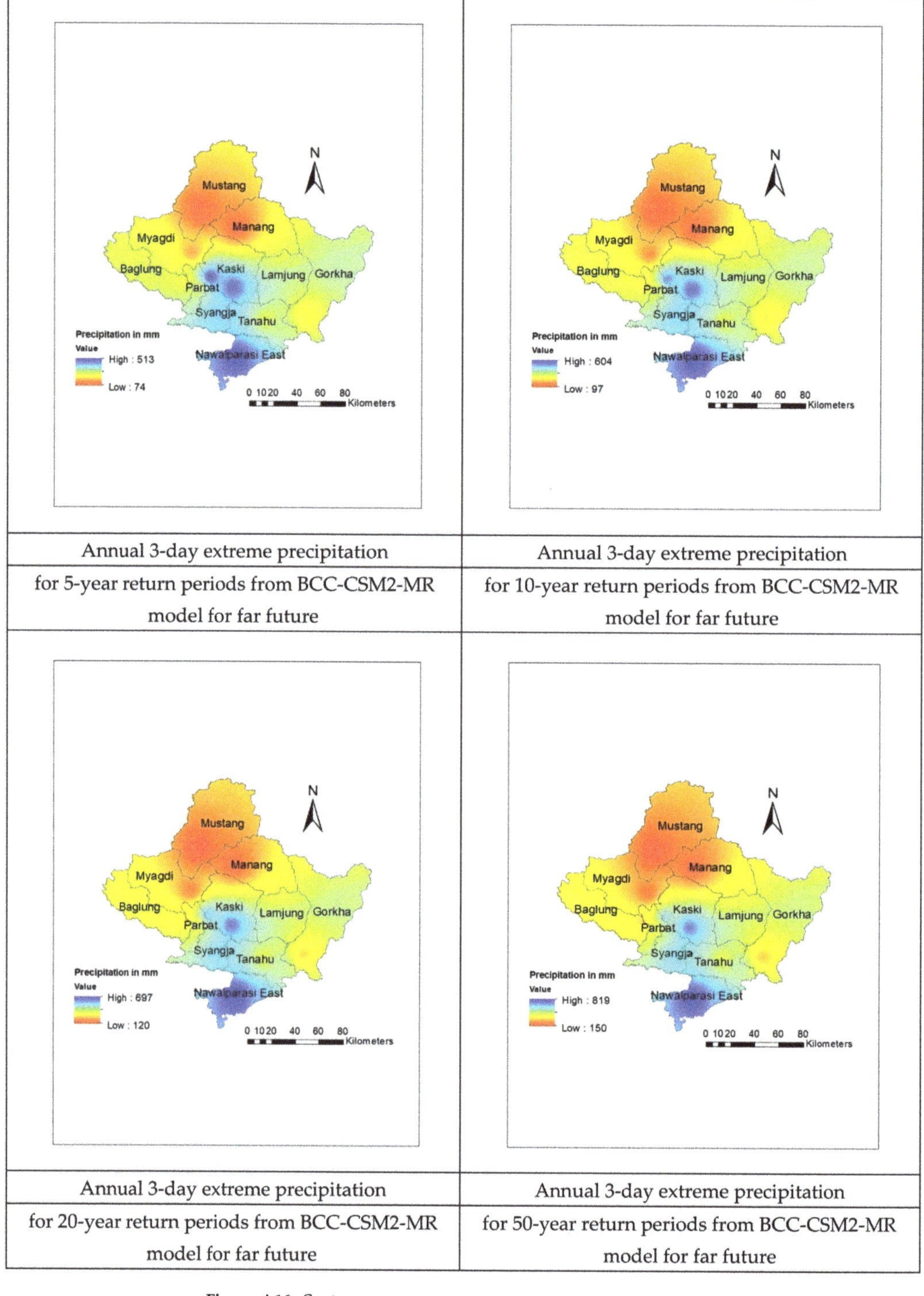

Annual 3-day extreme precipitation for 5-year return periods from BCC-CSM2-MR model for far future	Annual 3-day extreme precipitation for 10-year return periods from BCC-CSM2-MR model for far future
Annual 3-day extreme precipitation for 20-year return periods from BCC-CSM2-MR model for far future	Annual 3-day extreme precipitation for 50-year return periods from BCC-CSM2-MR model for far future

Figure A16. *Cont.*

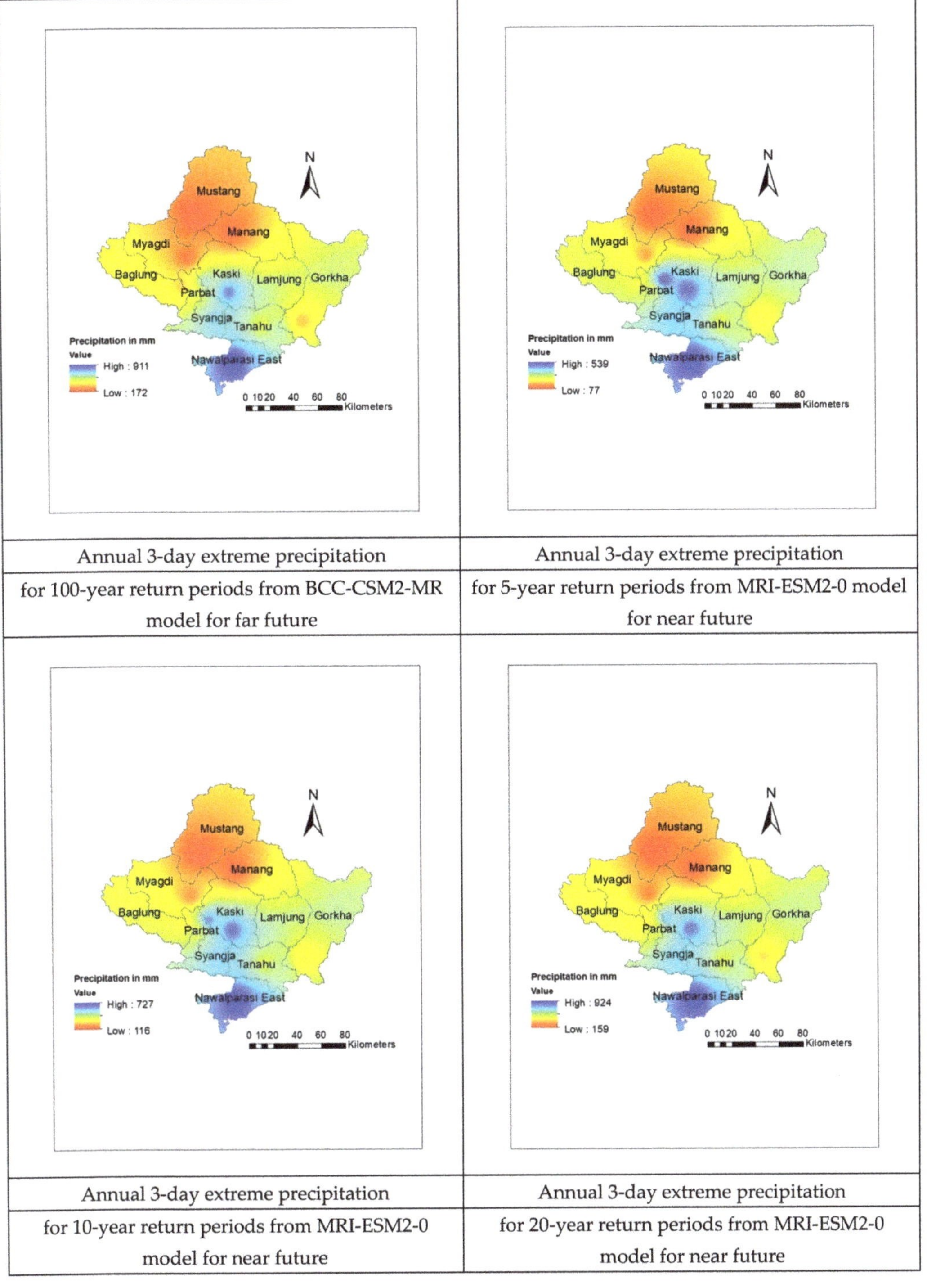

Annual 3-day extreme precipitation for 100-year return periods from BCC-CSM2-MR model for far future	Annual 3-day extreme precipitation for 5-year return periods from MRI-ESM2-0 model for near future
Annual 3-day extreme precipitation for 10-year return periods from MRI-ESM2-0 model for near future	Annual 3-day extreme precipitation for 20-year return periods from MRI-ESM2-0 model for near future

Figure A16. *Cont.*

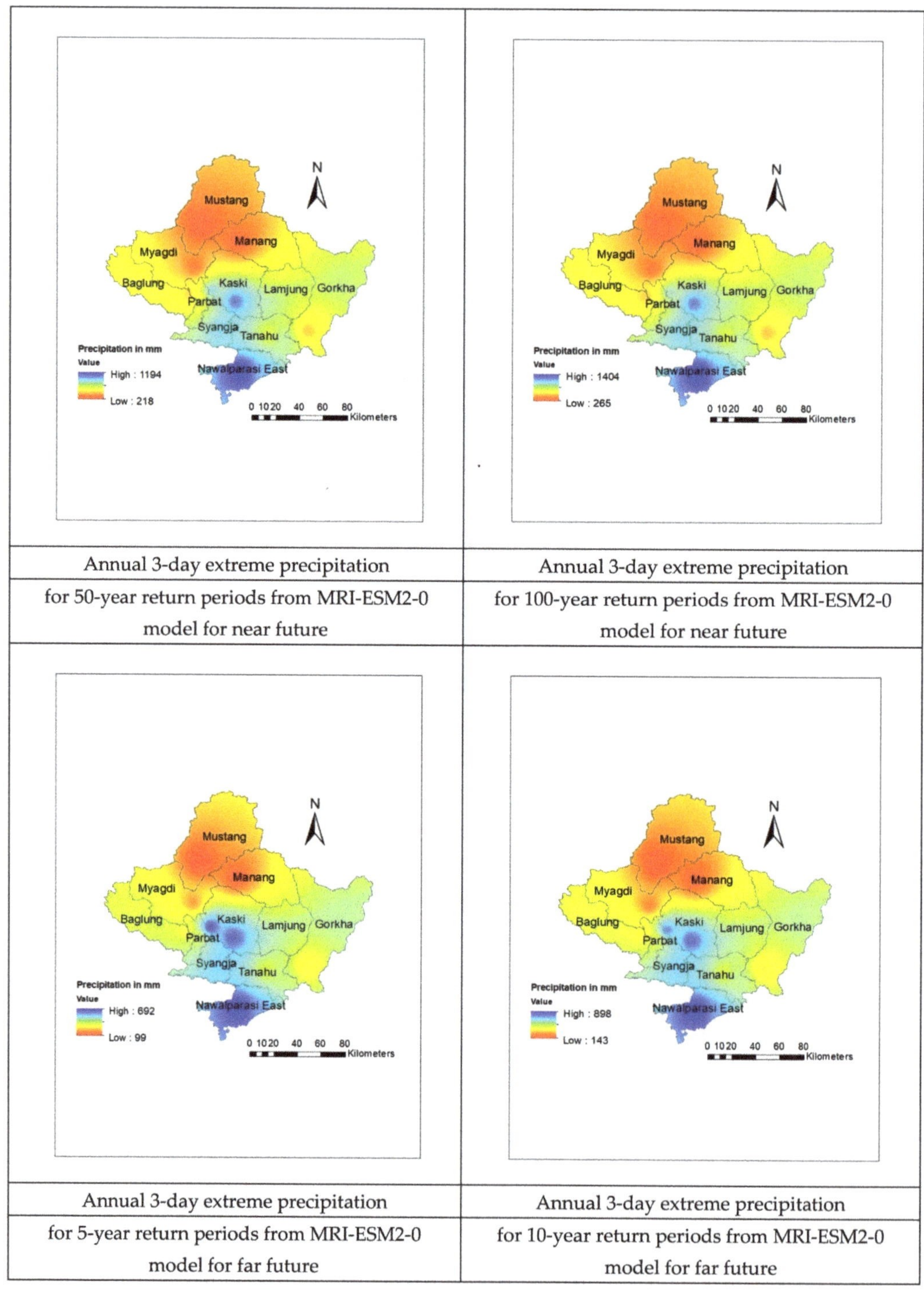

Annual 3-day extreme precipitation for 50-year return periods from MRI-ESM2-0 model for near future	Annual 3-day extreme precipitation for 100-year return periods from MRI-ESM2-0 model for near future
Annual 3-day extreme precipitation for 5-year return periods from MRI-ESM2-0 model for far future	Annual 3-day extreme precipitation for 10-year return periods from MRI-ESM2-0 model for far future

Figure A16. *Cont.*

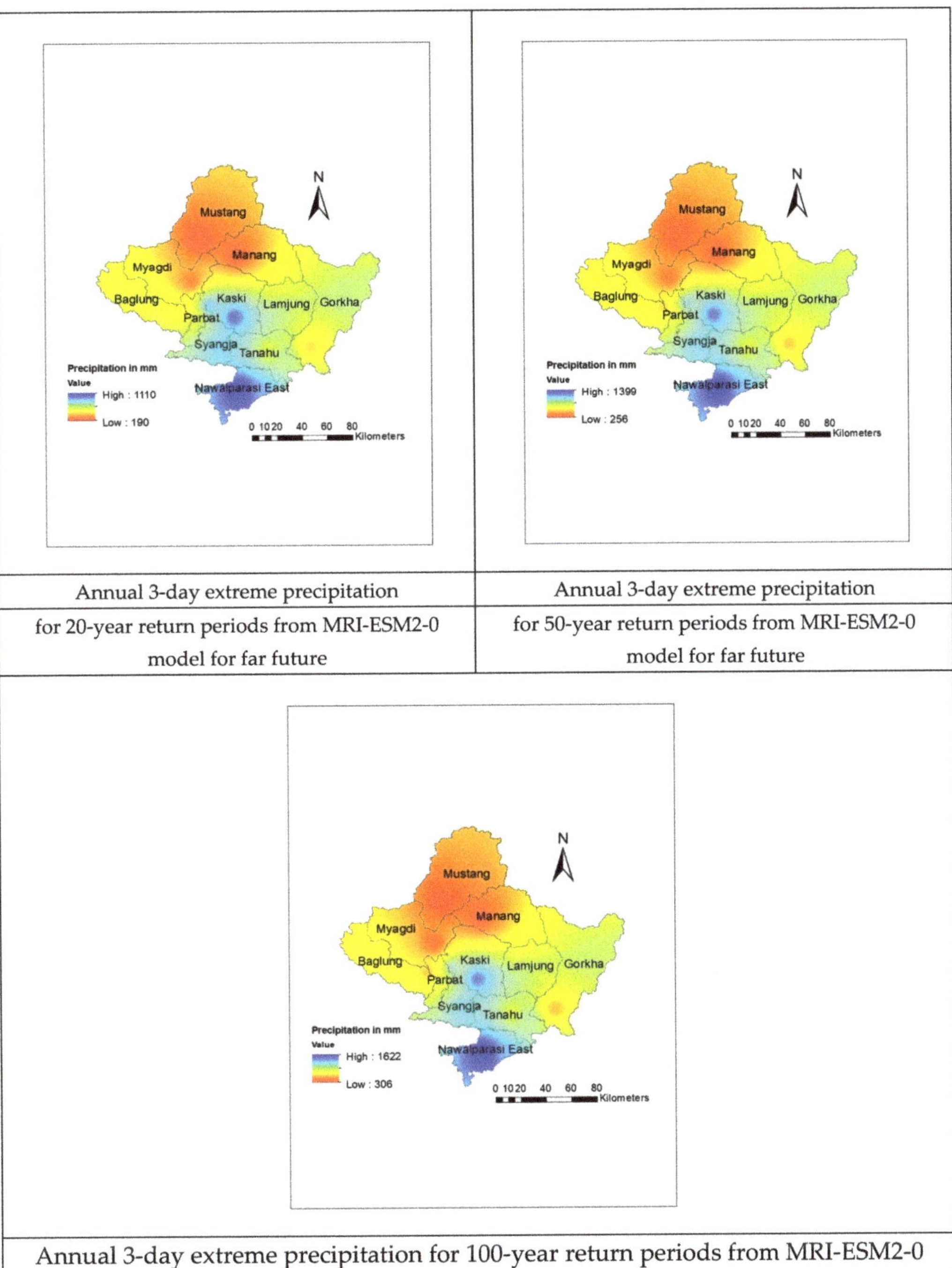

Annual 3-day extreme precipitation for 20-year return periods from MRI-ESM2-0 model for far future	Annual 3-day extreme precipitation for 50-year return periods from MRI-ESM2-0 model for far future

Annual 3-day extreme precipitation for 100-year return periods from MRI-ESM2-0 model for far future

Figure A16. Projected spatial annual 3-day extreme precipitation.

References

1. Basnet, K.; Shrestha, A.; Joshi, P.C.; Pokharel, N. Analysis of Climate Change Trend in the Lower Kaski District of Nepal. *Himal. J. Appl. Sci. Eng.* **2020**, *1*, 11–22. [CrossRef]
2. Chikosi, E.S.; Mugambiwa, S.S.; Tirivangasi, H.M.; Rankoana, S.A. Climate change and variability perceptions in Ga-Dikgale community in Limpopo Province, South Africa. *Int. J. Clim. Chang. Strateg. Manag.* **2019**, *11*, 392–405. [CrossRef]
3. Mansell, M.G. The effect of climate change on rainfall trends and flooding risk in the West of Scotland. *Nord. Hydrol.* **1997**, *28*, 37–50. [CrossRef]
4. Karki, R.; Hasson, S.U.; Schickhoff, U.; Scholten, T.; Böhner, J. Rising precipitation extremes across Nepal. *Climate* **2017**, *5*, 4. [CrossRef]
5. Pariyar, S.K.; Keenlyside, N.; Sorteberg, A.; Spengler, T.; Bhatt, B.C.; Ogawa, F. Factors affecting extreme rainfall events in the South Pacific. *Weather Clim. Extrem.* **2020**, *29*, 100262. [CrossRef]
6. Nayava, J.L. Heavy Monsoon Rainfall in Nepal. *Weather* **1974**, *29*, 443–450. [CrossRef]
7. Shrestha, A.B.; Wake, C.P.; Mayewski, P.A.; Dibb, J.E. Maximum temperature trends in the Himalaya and its vicinity: An analysis based on temperature records from Nepal for the period 1971–1994. *J. Clim.* **1999**, *12*, 2775–2786. [CrossRef]
8. Oruc, S. Investigation of The Effect of Climate Change on Extreme Precipitation: Tekirdağ Case. *Turk. J. Water Sci. Manag.* **2020**, *4*, 136–161. [CrossRef]
9. Ye, W.; Li, Y. A method of applying daily GCM outputs in assessing climate change impact on multiple day extreme precipitation for Brisbane River Catchment, MODSIM11. In Proceedings of the 19th International Congress on Modelling and Simulation, Canberra, Australia, 12–16 December 2011; Chan, F., Marinova, D., Anderssen, R.S., Eds.; Modelling and Simulation Society of Australia and New Zealand: Canberra, Australia, 2011; pp. 3678–3683.
10. Basnet, K.; Paudel, R.C.; Sherchan, B. Analysis of Watersheds in Gandaki Province, Nepal Using QGIS. *Tech. J.* **2019**, *1*, 16–28. [CrossRef]
11. Khadka, D.; Pathak, D. Climate change projection for the marsyangdi river basin, Nepal using statistical downscaling of GCM and its implications in geodisasters. *Geoenviron. Disasters* **2016**, *3*, 1–15. [CrossRef]
12. Mishra, B.K.; Mansoor, A.; Saraswat, C.; Gautam, A. Climate change adaptation through optimal stormwater capture measures. *APN Sci. Bull.* **2019**, *9*, 28–37. [CrossRef]
13. Mishra, B.K. Precipitation change assessment over upper Bagmati river basin using regional bias corrected GCM data. *Int. J. Water* **2017**, *11*, 294–313. [CrossRef]
14. Martel, J.L.; Mailhot, A.; Brissette, F.; Caya, D. Role of natural climate variability in the detection of anthropogenic climate change signal for mean and extreme precipitation at local and regional scales. *J. Clim.* **2018**, *31*, 4241–4263. [CrossRef]
15. Ghimire, P. A Review of Studies on Climate Change in Nepal. *Geogr. Base* **2019**, *6*, 11–20. [CrossRef]
16. Ayugi, B.; Tan, G.; Ruoyun, N.; Babaousmail, H.; Ojara, M.; Wido, H.; Mumo, L.; Ngoma, N.H.; Nooni, I.K.; Ongoma, V. Quantile mapping bias correction on rossby centre regional climate models for precipitation analysis over Kenya, East Africa. *Water* **2020**, *12*, 801. [CrossRef]
17. Ouali, D.; Cannon, A.J. Estimation of rainfall intensity–duration–frequency curves at ungauged locations using quantile regression methods. *Stoch. Environ. Res. Risk Assess.* **2018**, *32*, 2821–2836. [CrossRef]
18. Baidya, S.K.; Shrestha, M.L.; Sheikh, M.M. Trends in Daily Climatic Extremes of Temperature and Precipitation in Nepal. *J. Hydrol. Meteorol.* **2008**, *5*, 47–50.

Review

A Study of Urban Planning in Tsunami-Prone Areas of Sri Lanka

U. T. G. Perera [1,*], Chandula De Zoysa [2], A. A. S. E. Abeysinghe [3], Richard Haigh [1], Dilanthi Amaratunga [1] and Ranjith Dissanayake [3]

[1] Global Disaster Resilience Centre, Department of Biological and Geographical Sciences, School of Applied Sciences, University of Huddersfield, Queensgate, Huddersfield HD1 3DH, UK

[2] Department of Civil Engineering, University of Moratuwa, Bandaranayake Mawatha, Moratuwa 10400, Sri Lanka

[3] Department of Civil Engineering, University of Peradeniya, Galaha 20400, Sri Lanka

* Correspondence: thisara.uduwarage@hud.ac.uk

Abstract: Tsunamis pose significant challenges for disaster reduction efforts due to the multi-hazard, cascading nature of these events, including a range of different potential triggering and consequential hazards. Although infrequent, they have the potential to cause devastating human and economic losses. Effective urban planning has been recognised as an important strategy for reducing disaster risk in cities. However, there have been limited studies on urban planning for tsunami-prone areas, and there have been wide ranging strategies adopted globally. This is an international study aimed at exploring the status of urban planning in tsunami areas and better understanding potential urban planning strategies to reduce disaster risk in coastal regions. Drawing upon the work of an international collaborative research team, in this article, we present the findings of a systematic review of the urban planning literature. Using the PRISMA guidelines, 56 papers were selected, and three guiding questions informed the review. Further empirical investigations were carried out in Sri Lanka by a local research team, including twelve semi-structured interviews with representatives from agencies in urban planning, construction, and disaster management, and a focus group representing town and country planning, architecture, structural engineering, disaster management, landscape and geospatial planning, building services, green buildings and infrastructure and environmental management fields. The combined analysis reveals insights into the characteristics of the literature, as well as the nature of existing strategies for urban planning in tsunami-prone areas, grouped into six broad themes: community participation, spatial planning, soft and hard engineering;,evacuation planning, and resilience thinking. The findings also reveal limitations in existing strategies, including their failure to address multi-hazard threats and systemic risk, as well as inadequate community participation, and limited access to timely disaster risk information. The findings are used to inform an initial model of urban planning strategies in tsunami-prone areas that can be used before a hazard event occurs, during and in the immediate response to a hazard event, and during recovery and reconstruction following a disaster.

Keywords: urban planning; tsunami; Sri Lanka; multi-hazard

Citation: Perera, U.T.G.; De Zoysa, C.; Abeysinghe, A.A.S.E.; Haigh, R.; Amaratunga, D.; Dissanayake, R. A Study of Urban Planning in Tsunami-Prone Areas of Sri Lanka. *Architecture* **2022**, *2*, 562–592. https://doi.org/10.3390/architecture2030031

Academic Editors: Iftekhar Ahmed, Masa Noguchi, David O'Brien, Chris Tucker and Mittul Vahanvati

Received: 15 May 2022
Accepted: 16 August 2022
Published: 23 August 2022

Publisher's Note: MDPI stays neutral with regard to jurisdictional claims in published maps and institutional affiliations.

1. Introduction

Tsunamis pose significant challenges for disaster reduction efforts due to the multi-hazard, cascading nature of these events, including a range of different potential triggering and consequential hazards. Although infrequent, they have the potential to cause devastating human and economic losses.

Significant international and national efforts have gone into developing tsunami early warning systems, especially for the Pacific and Indian Ocean regions. However, recent tsunami events have illustrated the limitations of these systems, especially for near field events that may offer a short mobilisation period for a tsunami threat to be detected and warning information to be issued to people at risk, and thereby, enable effective evacuation and response efforts.

Effective urban planning has been recognised as an important strategy for reducing disaster risk in cities. However, there have been limited studies on urban planning for tsunami-prone areas and there have been wide ranging strategies adopted globally. Many also fail to consider the complexity of multi-hazard threats.

An international study was undertaken to explore the status of urban planning in tsunami-prone areas in order to better understand potential strategies to reduce disaster risk in these coastal regions. Drawing upon the work of a collaborative research team, in this article, we present the findings of a systematic review of the international urban planning literature on tsunami-prone areas and a detailed qualitative study in Sri Lanka.

Sri Lanka is a member state of the Indian Ocean Tsunami Warning and Mitigation System (IOTWMS) and was severely impacted by the 2004 Indian Ocean Tsunami. Despite only experiencing a single devastating tsunami event in modern history, a tsunami has been identified as the highest risk index among all hazard types in the country (8.9 out of 10) [1]. As highlighted by Haigh et al. [2], prior to 2004, there was little in the way of formal preparedness for tsunami early warning in the country and there was a lack of experience among the people, increasing their vulnerability. After 2004, there has been widespread recognition of a need for Sri Lanka to actively engage in disaster risk reduction efforts because of its vulnerability to future tsunami threats [3]. These investments in mitigation and preparedness for tsunami make Sri Lanka a useful focus for further detailed enquiry and to examine the extent to which urban planning features among those efforts. Three guiding research questions were used to scope the data collection and analysis.

1.1. Coastal Cities and Disaster Risk

The world has faced an era of exponential growth in its urban population. Cities are already home to more than half of the world's population, and are centres of economic growth and innovation [4]. Many of these cities are also situated in hazard-prone areas—along coasts or on floodplains, on top of or near seismic faults, in the shadow of volcanoes, and in areas prone to tropical cyclones and extreme storms. These natural hazards tend to exert even greater impacts since cities concentrate people and economic assets, including large-scale infrastructure and properties [4]. Specific threats include rapid-onset hazards such as major cyclones accompanied by high winds, waves and surges, or tsunamis. They are also threatened by slow-onset hazards, including erosion and gradual inundation [5]. The effects of climate change can be especially devastating to vulnerable coastal areas, including the function and structure of their ecosystems. Sea level rise changes the shape of coastlines, and contributes to the risk of coastal erosion, flooding, and salt-water intrusion. The UNFCC [6] estimates that more than 600 million people (around 10 per cent of the world's population) live in coastal areas that are less than 10 m above sea level, and these occupy 7% of the Earth's land area.

1.2. Tsunami and Its Complexity as a Cascading Disaster

Tsunamis are typically triggered by earthquakes, but can also be triggered by other hazards, including volcanic eruptions, submarine landslides, and by onshore landslides in which large volumes of debris fall into the water. Although less frequent than most other natural hazards, tsunamis have the potential to cause devastating property damage and loss of life. The International Tsunami Information Centre reports that over the last twenty years, deadly tsunamis have been triggered near Chile (2007, 2010), Haiti (2010), Indonesia (2004, 2005, 2006, 2010, 2018 (two separate events)), Japan (2011), Peru (2001), Samoa-American Samoa-Tonga (2009, 2022), and the Solomons (2007). Although such devastating events have been rare, some recent studies have suggested that even minor sea level rise will increase the risk of tsunamis for coastal communities [7]. Similarly, the growth of urban populations in coastal regions is likely to increase exposure of people and assets to the threat posed by tsunamis.

The 2004 Indian Ocean Tsunami, which impacted twelve countries, is one of the worst disasters in recorded human history. The 2011 Tohoku Earthquake and Tsunami event, often referred to as the Great East Japan Earthquake and Tsunami, caused the sea water to reach more than 5 km inland in some coastal locations, reaching a height of 19.5 m above sea level. In both events, larger numbers of people lost their lives and were displaced, losing their homes and property. [8–10].

In late 2018, Indonesia was hit by two further, destructive tsunamis, which as reported by IOC [10], challenged the traditional understanding of tsunami hazard, warning, and response mechanisms. At an International Symposium on the Lessons Learnt from the 2018 Tsunamis in Palu and Sunda Strait [11], experts identified a need for a critical dialogue on the future direction of the Indian Ocean Tsunami Warning and Mitigation System (IOTWMS), especially for events other than tectonic origins and with short warning times, referred to as local source or near field tsunami. They concluded that the national warnings issued in Indonesia within 5 min of the earthquake, were of limited practical use for the Palu tsunami, especially in coastal areas where tsunami waves arrived in less than 3 min. They also determined that current early warning systems are most effective for tsunamis generated by subduction zone earthquakes but have limitations to handle atypical (landslide and volcanic) and/or near-field tsunamis.

Tsunamis such as the one in Palu are often associated with multi-hazard threats with cascading effects. Goda et al. [12] explained that the cascading effects of the 2018 Sulawesi earthquake were very complex and varied widely in space, which triggered secondary hazards, such as a tsunami, liquefaction, submarine landslides, and massive mudflows in coastal areas. Muhari [13] added that such cascading effects could increase in progression over time and generate unexpected secondary events of strong impact across the coastal regions.

The 2011 Tohoku Earthquake and Tsunami event also illustrated the potential devastation caused by cascading effects of earthquakes and tsunamis [14], in this case, damage to the local nuclear power plant caused extensive radioactive contamination. It also caused widespread damage to critical infrastructure, including transport such as roads, 71 bridges, and 26 parts of the railway system, and lifeline infrastructure such as electricity, water supply, sewage systems, and gas lines.

These previous disasters illustrate the complexity facing tsunami early warning and preparedness efforts, which must overcome the dilemma of time verses uncertainty due to short response times, limitations of technology, and currently available scientific knowledge. There has been considerable progress and improvement in the IOTWMS that has been developed since the devastating 2004 Indian Ocean Tsunami. However, the UN coordinating agency for the IOTWMS has formally recognised that much remains to be done to ensure dissemination of effective warnings and to prepare communities to act upon them [11,15]. Most recently, recommendations from the Lessons Learnt Symposium [11] stressed a need to address the lack of proper evacuation plans and related infrastructure during a tsunami emergency, as well as poorly implemented/ineffective spatial planning and policy-related issues.

Poorly planned urban environments, weak urban governance, and old and fragile infrastructure can all increase pressure on the urban environment and trigger exposure to these cascading effects. They are associated more with the magnitude of vulnerability of the coastal regions than with that of the hazard. As highlighted by Rus [16], the need for maintenance and upkeep of these cities makes safety measures for their citizens crucial in these situations. Therefore, urban planning has been recognised as a key factor to build the resilience of urban coastal communities in the face of increasing hazards and disaster risk.

1.3. Urban Planning and Disaster Risk

Urban planning encompasses the preparation of plans for and the regulation and management of towns, cities, and metropolitan regions. It attempts to organize socio-spatial relations across different scales of government and governance, and is concerned with the social, economic, and environmental consequences of delineating spatial boundaries and influencing spatial distributions of resources [17]. In addition, it focuses on attempting to improve the plans, functions, and management of cities and areas, which has a clear role to play in disaster mitigation [18].

Urban planning can serve as a key tool for limiting the disastrous effects of hazards and for enhancing community resilience. This has been emphasized in recent international agreements, i.e., the Hyogo Framework for Action (2005) [19], the Sendai Framework for Disaster Risk Reduction (2015) [20], and the 2030 Agenda for Sustainable Development (2015) [21], which highlight the importance of incorporating disaster risk reduction and resilience enhancement into urban and regional planning processes [22].

Alan [23] also highlighted the potential to reduce disaster risk through proper urban planning approaches, where understandings of risk are increasingly being integrated within the wider processes of urban planning. For example, proper planning allows cities to control land use and occupation, which is a crucial factor in minimising exposure to a tsunami and flooding. Many recent studies have also highlighted the importance of land use planning in coastal regions as a crucial factor for urban resilience [16,24,25].

Some countries have used a variety of urban planning instruments to monitor land use and minimize the effects of specific threats such as tsunamis. These include resettlement, the development of no-build zones, land use restrictions, and tsunami-safe building standards, as well as the preservation of natural buffers including forests, compacted dunes, and wetland areas [26]. Furthermore, Eisenman [27] highlighted the importance of non-statutory policies such as evacuation, mandatory insurance, emergency, and the promotion of social organizations on community resilience, which can be promoted through effective urban planning.

Although some of these measures may be suitable for dealing with tsunamis, they may not be appropriate measures for other types of hazards or threats, for instance, seismic activity, severe storms, or landslides. Bosher [28] pointed out that multi-hazard/threat assessments should be undertaken, and any risk reduction options should be proportionately considered alongside any other hazards or threats that have been identified. Despite this recognition of the need, Ma et al. [29] found that the vulnerability of a city or area to hazards was not being adequately evaluated, especially in developing countries and even where geo-hazards were likely to occur from time to time due to the poor physical environment and strong human activities. Sengezer and Koç [30] also revealed that a high risk of exposure to multiple hazards in such areas is associated with considerable deficiencies in the social and political spheres. This includes a lack of professional training programmes, poor access to technology, lack of experience of organisations, inefficient professional chambers, and inadequate control mechanisms. These poor mechanisms have the potential to do serious harm to people's lives and property in coastal regions of the world [31].

The "new normal" that is expected to emerge as the world recovers from COVID-19 is also likely to further change the complexity of disaster risk, as urban areas change to reflect societal changes, such as remote working [32,33]. Urban planning will need to evolve so that it can continue to contribute to mitigation and preparedness efforts. A better understanding of the trends and dynamics of pandemics, and their impact on urban planning for coastal cities, response, and adaptation measures, will be vital in tackling this dynamic situation.

Although a large body of research has been published on various strategies related to urban planning in coastal areas, there appears to be a lack of consensus or best practice around what measures to adopt in tsunami-prone areas. The combined effects and cascading effects of hazards are also not addressed in most of the existing research.

2. Materials and Methods

This study aims to explore the state of the art in urban planning for tsunami-prone areas and to better understand good practices and gaps in the knowledge base. Through a series of meetings among an international research team from Sri Lanka and the UK, the following research questions were developed to guide the study:

RQ1 What are the characteristics of the literature on urban planning in tsunami-prone areas?

RQ2 What are the existing strategies and approaches for urban planning in tsunami-prone areas?

RQ3 What are the main gaps and limitations in those strategies?

A systematic review of the literature was carried out by the UK research team on urban planning strategies in tsunami-prone areas. The Preferred Reporting Items of Systematic Reviews and Meta-Analysis guidelines (PRISMA) were used to prepare and facilitate transparent reporting of the systematic review [34]. This guideline enables readers to assess the quality and validity of the reporting. After the literature search, a frequency analysis was undertaken to identify and categorise the applied methods from literature to better frame the urban planning strategies.

The search terms "(Urban Planning* AND (Tsunami) AND/OR (Multi Hazards)" were used in Summon and University of Huddersfield Library Services, Scopus, and Web of Science. These research services were used to reach a wide range of research across multiple databases. These publication platforms reliably search across publishers and are not bias towards journals published by any one company. In total, the search methods identified 1918 records. The year range of 2000–2020 was applied to focus on the considerable body of literature that had emerged since the 2004 Indian Ocean Tsunami and to ensure that it reflected the most recent studies under the urban planning literature. After searching all the databases, the studies were added to EndNote software and the duplicate studies were deleted. Further, the manual search method was applied in the EndNote software to find duplicates. Then, the titles and summaries were reviewed to find relevant studies. Subsequently, experienced scholars specializing in the field independently reviewed the full text of the articles and applied the following inclusion criteria:

1. Research papers, book chapters, journal articles are considered;
2. Research should include the term/terms in the title abstract or the keywords;
3. It should be in English;
4. Research should be related to one or more domains of urban planning strategies in tsunami-prone areas.

The following exclusion criteria were also applied:

1. Urban planning is not a main focus;
2. Does not include tsunamis or other coastal hazards;
3. Does not include urban planning strategies or mechanisms;
4. Does not address research questions of the study.

After applying the inclusion and exclusion criteria, 56 academic research papers (Appendix A Table A1) were selected and reviewed to address the research questions (Figure 1).

In addition to the systematic review, empirical investigations were planned and carried out in Sri Lanka by a local research team but using a methodological approach jointly developed by the UK and Sri Lanka team. These contemporary insights into current urban planning for tsunami-prone areas in Sri Lanka, including its current strategies, foci, and limitations, complement and contextualize the findings from the systematic review, which provides much broader insights.

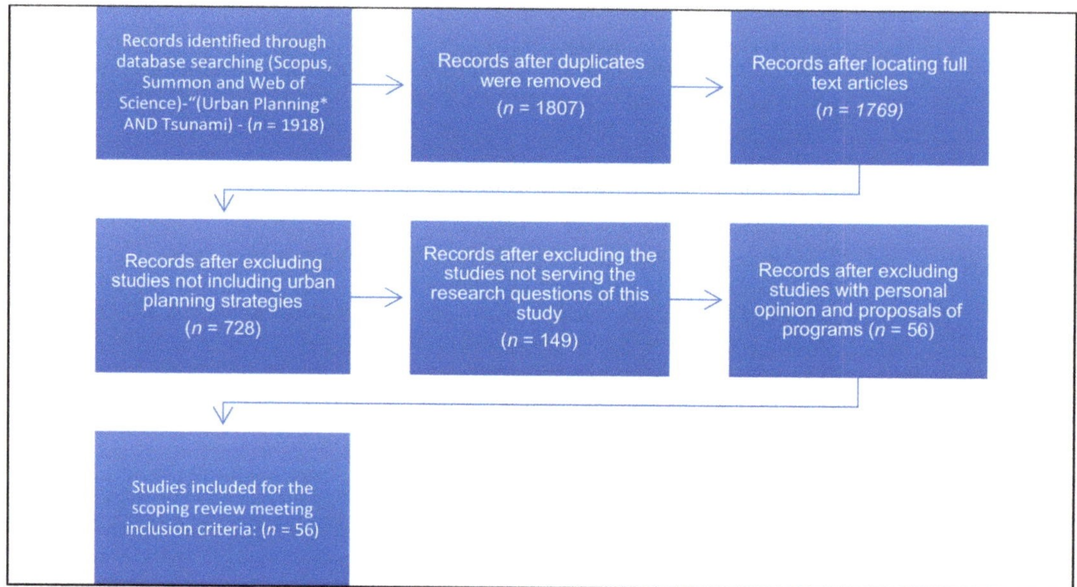

Figure 1. Flow diagram of the systematic literature review.

The empirical investigations involved semi-structured expert interviews and a focus group to investigate the existing strategies and approaches for urban planning in tsunami-prone areas of Sri Lanka, systemic risk and cascading impacts, and the constraints and pathways to mainstreaming disaster risk reduction into urban planning at the national to local levels. This dual approach combined the benefits afforded by the detail and individual contexts from interviews, as well as the wider range of ideas and greater "surface" data from a focus group [35]. It also enabled convergence of the current status, good practices, and gaps across focus groups and individual interviews, which enhanced the trustworthiness of the findings.

Twelve interviews were conducted in Sri Lanka with representatives from agencies in urban planning, construction, and disaster management (Table 1). These interviews were an opportunity to explore beyond the literature and allow additional concepts, themes, and areas of analysis to be discovered. Further, the focus group in Sri Lanka was used to examine further and validate the findings from the interviews, while also providing an opportunity to clarify issues and ensure nothing significant was overlooked. Six experts took part in a one and half hour focus group discussion, representing town and country planning, architecture, structural engineering, disaster management, landscape and geospatial planning, building services, green buildings and infrastructure, and environmental management fields. An audio recording of the focus group discussion was analysed using content analysis.

Data collection for the systematic review and interviews/focus group was conducted concurrently and largely independently. Then, both strands of research were combined during data analysis and, primarily, interpretation. Findings from the systematic review were cross-referenced, triangulated, and contextualized with expert observations to gain a deeper understanding of urban planning strategies in tsunami-prone areas within the theoretical discourse, research gaps, and potential discrepancies, with observations from practice.

Table 1. Interview respondent expertise.

Expert Area	Code
Architecture, Town Planning	A1
Architecture, Town Planning	A2
Architecture, Town Planning	A3
Civil Engineering	C1
Civil Engineering, Disaster Management	C2
Civil Engineering, Disaster Management, Sustainable Built Environment	C3
Civil Engineering	C4
Disaster Management	D1
Disaster Management	D2
Green Building Consultant, Town Planning	G1
Town and Country Planning	T1
Town and City Planning	T2

Data Analysis

A frequency analysis was conducted to record the major concerns and dimensions for urban planning strategies found through the systematic review, as well as the aspects of the urban planning applied in the literature. In the review, the data was organised by the following categories: case information, characteristics of urban planning, dimensions of urban planning, and applied methods. The case information included source, types of hazard described in the study (topic), the spatial scale, and the type of territory where the assessment methods applied. The spatial scale was further categorized by the size at which the assessment was operated. It ranged from the community, district, city, region, country to the global scale. Further analysis for the systematic review included expert interviews and a focus group that were developed based on a thematic extraction, where key themes of each research and interviews were extracted. Then, the themes were classified based on the research questions, and these key findings of the classification are highlighted in the results and discussion section. Based on the above analysis, the key strategies were derived to form a conceptual framework related to urban planning in tsunami-prone areas.

3. Results

3.1. Characteristics of the Literature

3.1.1. Location of the Research

Figure 2 illustrates the geographical location of the selected research studies in the systematic review. The Indian Ocean tsunami in 2004 appears to have prompted a substantial body of studies related to urban planning, with several countries in the region strongly represented. Indonesia, which experienced the highest death toll in the disaster, has been the most frequent target of studies, with 11 of the 56 addressing the country. Indonesia has also experienced very high levels of population and economic growth, and has urbanized areas spread across 6000 inhabited islands, and thereby, faces major challenges in terms of urban development. Sri Lanka ($n = 7$), the country that experienced the second highest death toll from the 2004 tsunami, also features prominently in the studies, while there are also a smaller number of studies in other affected countries, including India ($n = 4$), Bangladesh ($n = 1$), Malaysia ($n = 1$), Thailand ($n = 1$), and the Maldives ($n = 1$). Several Pacific Ocean countries also feature prominently, including Japan ($n = 5$), which suffered high human and economic losses from the 2011 Tohoku Earthquake and Tsunami, and Chile ($n = 7$), which experienced a tsunami following a magnitude 8.8 earthquake in 2010.

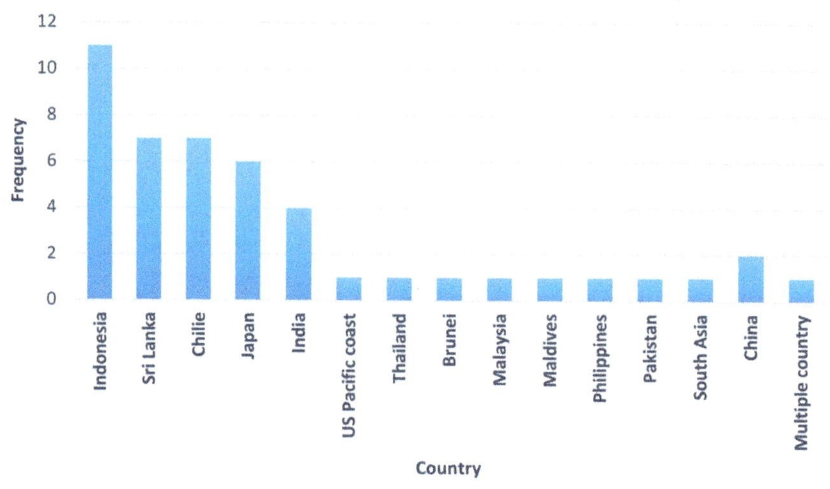

Figure 2. Study area of the selected sources.

3.1.2. Spatial Scale

Figure 3 illustrates the spatial scale of the reviewed sources in the systematic review. City level and country level scales were the prominent focus in most of the research. These two scales covered more than half of the reviewed sources (60%). Another characteristic from the literature included in this study is that the majority of research has been concentrated on larger cities (such as Jakarta, Aceh, and Padang), whereas the latter (smaller cities and towns) have been far less explored and therefore understood. This is significant because smaller cities and towns are likely to have different features and urban planning issues than bigger, major cities. Smaller cities and villages, for example, are likely to have less decentralised power and resources than capital cities, which may have an influence on local action in urban planning response to catastrophe risk [36].

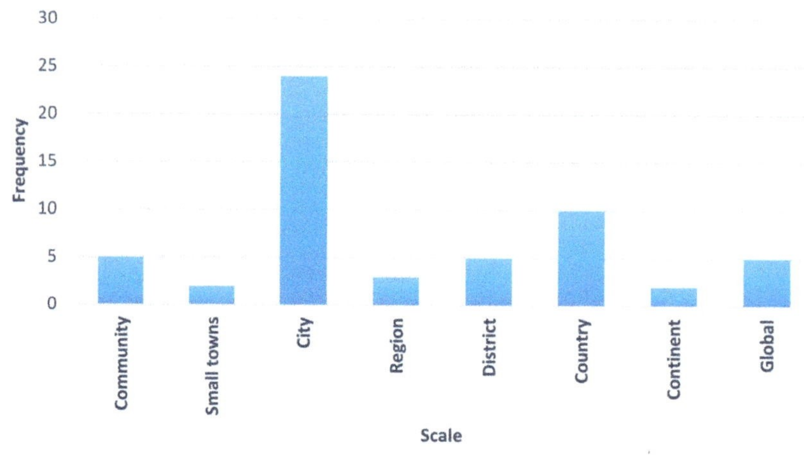

Figure 3. Spatial scale of the selected sources.

3.1.3. Year of Publication

Figure 4 sets out the frequency of included studies against the year of publication. A substantial increase in the number of publications can be observed from 2013. In short, the majority of included studies were published in the last six to seven years. This could be due to an increase in awareness about urban planning and their role in tackling disaster risk, or an overall increase in disaster risk and major losses linked to natural hazards. It could also be an indication that this is an area of research that is growing in interest, with more work being published. Alternatively, it may also reflect an increase in the number of disaster risk related journals and overall publication volumes in recent years. Regardless, the implications of that this is an area of increasing interest, and it is very likely that in the next few years there will be more relevant literature in this area.

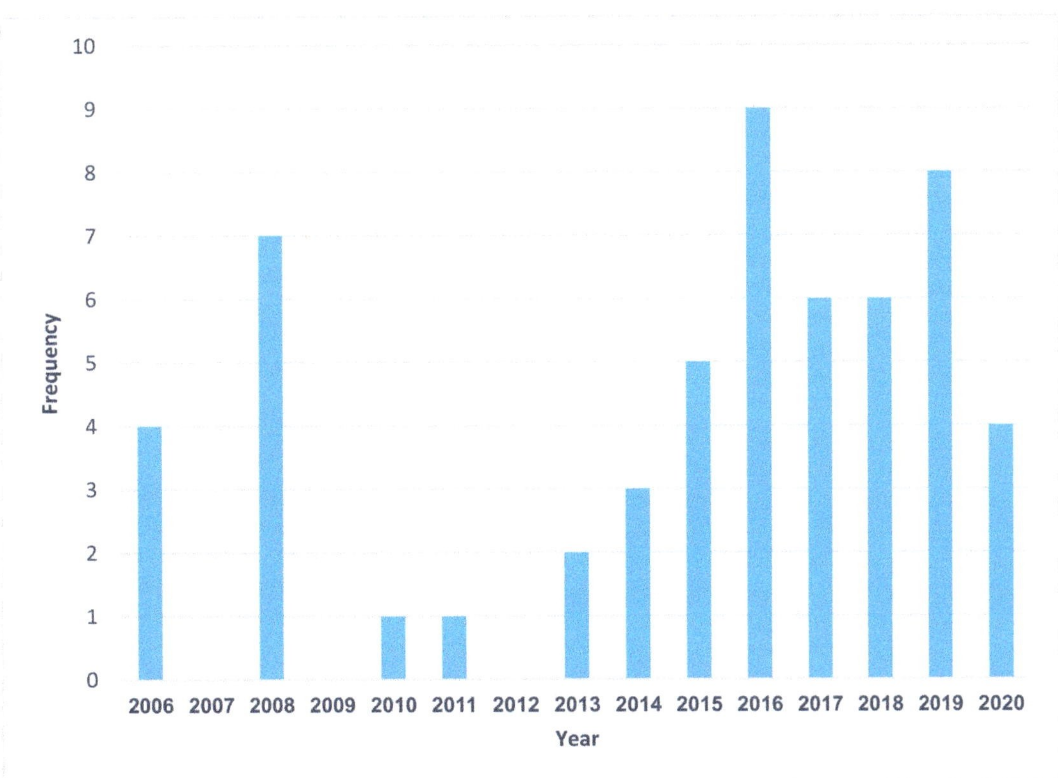

Figure 4. Published year of the selected sources.

3.1.4. Research Methods

In terms of research methodology and as represented in Figure 5, a wide range of different approaches were adopted across the studies. A general observation was the predominance of studies utilising qualitative approaches, most frequently based on perception data such as a limited number of purposely sampled interviews, literature, or secondary data. Despite all selected articles being in academic peer-reviewed journals, several had no clearly articulated method. Even those that did, tended not to describe data collection techniques at length, and many did not describe detailed data analysis approaches. An in-depth analysis of the quality of quality of evidence, such as weight of evidence, was not carried out. However, it was observed that many used purposive samplings, had small sample sizes, or did not specify in sufficient detail.

The lack of common methodological approaches across studies, as well as the lack of detail in the presentation of research methods, makes direct comparisons across studies more difficult. It also does not lend itself to research synthesis, which could otherwise be used to inform urban planning in different contexts.

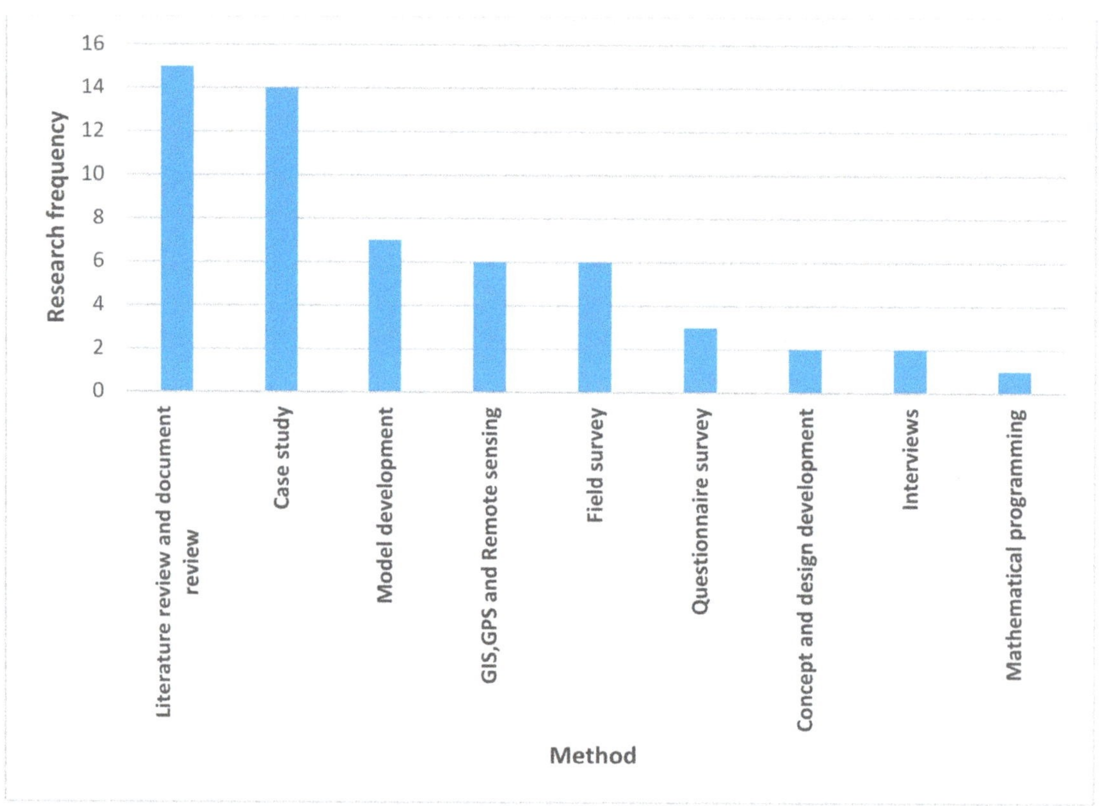

Figure 5. Research methods of the selected sources.

3.1.5. Other Hazards in Tsunami-Prone Areas

As the main focus of the study, tsunami hazards feature prominently in the selected studies, with over 30 either completely or partially investigating tsunami risk in coastal cities. As illustrated in Figure 6, other hazards in the selected papers include floods ($n = 7$), pandemics ($n = 2$), earthquakes ($n = 1$), and cyclones ($n = 1$). Figure 4 sets out the frequencies of studies included in this systematic review focusing on different urban hazards. Other hazards, such as coastal storms, wildfires, and landslides were not discussed in the urban planning literature. This resonates with Murray [36] who also noted in their review that many of the studies on urban risk focused on floods, earthquakes and tsunamis, whereas other hazards are underrepresented.

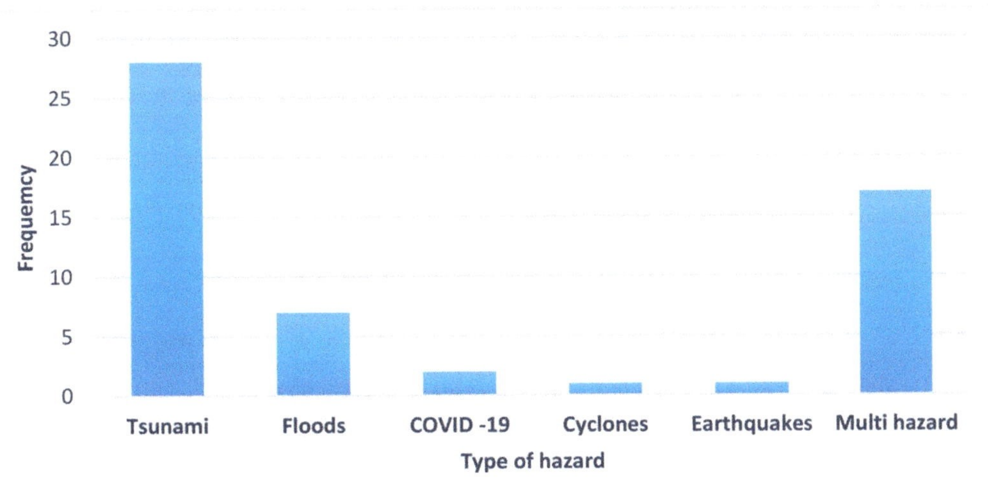

Figure 6. Frequency of hazards in tsunami-prone areas in selected sources.

Among the other hazards in tsunami-prone areas, floods in coastal regions were frequently highlighted as the prominent hazard, especially in South and South East Asia [37,38]. Dhiman [37] revealed that the risks posed by floods in coastal cities were primarily due to (1) the riverine flooding and extreme precipitation; (2) the effect of cyclonic storms, tsunamis, and tidal inundation; (3) failure in urban planning and unsustainable land use practices. The impact of earthquakes is also discussed in some of the papers combined with tsunamis, which pose a great threat to coastal cities as a stand-alone event or a tsunamigenic event. The 2004 Indian Ocean Tsunami, which was triggered by an earthquake with an epicentre off the west coast of Northern Sumatra, Indonesia, affected many coastal cities in Southeast and South Asia. Most of the reviewed papers highlighted the impact of the event, which also prompted planners to develop urban planning strategies to tackle similar events in the future [39,40]. Among the selected urban planning literature, there were very few studies on biological hazards, such as the COVID 19 pandemic and Ebola outbreak (n = 2). Due to the wide reach and devastating impact of the COVID-19 pandemic, it is reasonable to expect that more studies that address biological hazards will emerge soon.

A substantial number of studies identified in this review were multi-hazard (n = 17), focusing on more than one hazard and often the interactions between hazards. For example, tsunamis are often a significant secondary hazard following an earthquake, especially in coastal cities in countries such as Chile, Indonesia, and Japan, which are located in the most seismically prone region of the world, the Pacific "ring of fire" [41]. Similarly, coastal cyclones are also liable to cause floods where urban settlements spread in an unplanned way along the coast, such as found in Mumbai and Kolkata [42]. Barría [43] also highlighted that flood risks were often coupled with cyclones, storm surges, tides, and tsunamis across different geographical locations in a multi-hazard scenario. Most of the studies highlighted the importance of integrating multi-hazard scenarios in urban planning, and thereby, to provide a proper estimation of the risk level in the regions. Doerner [44] cautioned that there was a high probability for a combination impact of two or more hazards, such as tsunamis and earthquakes or floods and landslides in the same time frame. Similarly, Goda [12] stressed the high probability of cascading effects due to multi-hazard scenarios.

3.2. Existing Strategies for Urban Planning in Tsunami-Prone Areas

As illustrated in Figure 7, the literature describes a range of urban planning strategies that have been implemented in tsunami-prone areas that could contribute to disaster risk reduction. The most discussed strategies include evacuation planning (13), community participation (12), and land use planning (11). Less frequently discussed are resilience thinking (6); soft engineering (4); physical and social (3); sustainable housing, and physical and health planning (2 each); and hard engineering, urban physical growth planning, and urban governance (1 each).

Evacuation planning is the most frequently discussed strategy in the reviewed sources ($n = 13$). Most of the reviewed papers highlighted the importance of evacuation planning as an urban planning strategy for disaster risk reduction in coastal regions. Further, these reviewed sources have applied evacuation planning strategies in connection with a range of coastal hazards, including tsunamis, earthquakes, floods, and cyclones. Yossyafra [45] also contended that many governments have paid more attention to evacuation planning for their disaster risk reduction efforts.

Among the studies on evacuation planning, a number of different approaches emerge. An improved road infrastructure system ($n = 6$) is frequently discussed as a way to improve evacuation planning. Safe routes development with emergency shelters are also commonly discussed ($n = 6$). Public open spaces ($n = 2$), escape hill development ($n = 2$), and vertical evacuation ($n = 2$) are some of the other evacuation planning approaches that are suggested as part of urban planning.

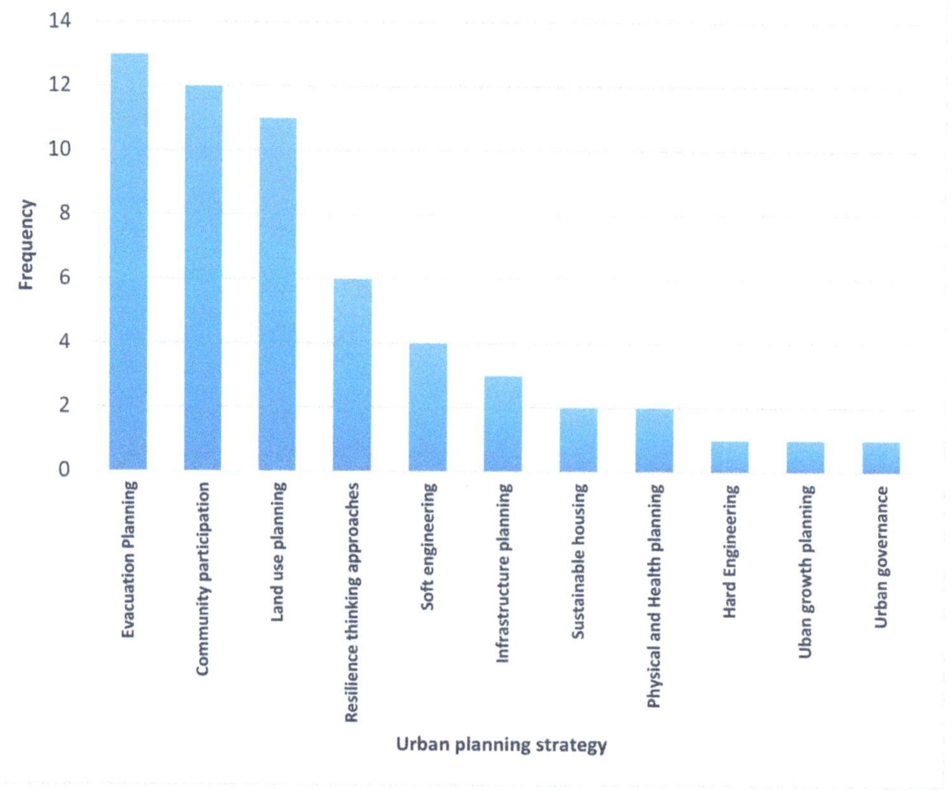

Figure 7. Existing strategies for urban planning in tsunami-prone areas.

Some studies have also suggested more innovative evacuation planning mechanisms, including personal life saving capsules [46], boat-based evacuation [47], and web-based interactive evacuation visualization tools [48]. These are suggested as ways to improve current evacuation planning strategies.

Most of the studies have focused on evacuation planning in the context of a single hazard event, rather than address any complications created by a multi-hazard scenario. This was also observed by Usha [49], in a study of coastal cities in India, who noted that flood evacuation mapping had been prepared without consideration of tsunamis and other coastal hazards.

Community participation was the second most discussed strategy ($n = 12$). Sustainable community engagement in the predisaster planning and post-disaster recovery processes were commonly cited themes. Many studies stressed the involvement of vulnerable communities. Spatial planning also featured prominently ($n = 10$). Among these studies, land use planning using digital technologies were the main focus. Nature-based planning ($n = 3$), structural planning ($n = 3$), and resilience thinking ($n = 3$) were among the other strategies discussed in the urban planning literature. These also tended to focus on single hazard events, rather than address multi-hazard scenarios.

Other social and economic aspects, including population growth planning ($n = 1$) and health-related planning ($n = 1$) have typically been ignored in the existing urban planning literature related to disaster risk reduction. However, the COVID-19 pandemic has again revealed the importance of health sector planning in these regions, where highly populated coastal cities were directly impacted.

As revealed by the reviewed sources, some of the urban planning approaches have become more prominent in recent years. These include urban resilience, population growth planning [50], and physical and social infrastructure planning [51,52]. These strategies tend to focus more on the demographical and social aspects in urban planning, for example, where the urban population growth must be controlled and managed effectively to achieve sustainability in these cities.

The following sections examine in more detail each of the main strategies (RQ2), as well as gaps and limitations that are revealed within the literature (RQ3).

3.2.1. Spatial Planning

Spatial planning is a main tool for urban planning that many governments use to influence the urban form and spatial development, including in tsunami-prone areas. The reviewed sources revealed land use planning ($n = 4$) as a major instrument for tackling disaster risk. Land use planning serves to demarcate zones or reserve land as buffers, which can make cities more resilient to the threats posed by a range of hazards [53]. Risk zone planning, which imposes restrictions for construction and development of infrastructure in specified zones, can be used to prepare coastal cities to cope with hazards such as tsunamis and coastal storms [54]. The literature also suggests that the inclusion of public open spaces and green spaces can be used to further empower the disaster resilience of cities while providing other ecosystem services for communities [55,56]. Similarly, Arif [57] suggested that the development of green open space, along with other zones such as commercial districts, residential areas, industrial areas, and transport networks, could influence the ecological balance as well as the economic balance of cities. Meutia [58] further discussed this concept with heritage site planning in a post-disaster context, where it could play an important role in the production of essential daily social functions, such as the provision of space for community consultation, conflict resolution, decision making, and disaster mitigation learning.

The reviewed sources show that governments can use land use planning strategies by empowering land use regulations through legislation and policy, and adopting planning instruments such as governmental statutes, regulations, rules, codes, and policies that can influence land use prior to a disaster event. For example, the relocation of residential developments in tsunami-prone zones or river floodplains, strongly decreases the vulnerability to a tsunami or river flooding [37]. In this sense, urban planning has strong potential for disaster risk reduction. Urban planning can set out land use regulations, including no-construction zones, and control development in hazard zones, and it can also define policies regarding evacuation, mandatory insurance, relocation of critical infrastructure, etc., which can notably reduce vulnerability to socio-natural disasters.

On a broad level, these planning instruments deal with the type, location, and extent of the land that is vulnerable to hazard events. Such vulnerability and risk assessments were frequently discussed in the reviewed sources ($n = 6$), especially with the application of digital technologies such as geographic information systems (GIS) and remote sensing. The development of GIS-based technologies and utilization of its tools for spatial planning were identified in several of the reviewed studies ($n = 4$). These technologies can be used to better understand hazards and to map vulnerability to reveal risk [49,59], which includes the physical and human factors such as topography, hydrography, land use, construction, settlements and communication networks in urban planning processes. Vulnerability assessments with digital technologies were also highlighted in the literature as an effective way to mitigate the impact of earthquakes and tsunamis [12,49,53]. Moreover, the utilization of satellite imagery in remote sensing and web-based risk assessment have provided new avenues for land use planning, which were highlighted in the studies as the next step of spatial planning for disaster risk reduction [59]. These technologies have been frequently used in tsunami and flood vulnerability mapping, while their use in connection with other hazards has been less evident.

3.2.2. Community and Key Stakeholder Participation

As with various types of spatial planning, consulting with the community and relevant stakeholders are an important part of land use planning to ensure transparency and incorporate a wide range of interests into the overall urban plan. This was highlighted in most of the studies, where communities and key stakeholder groups play significant roles in the urban planning process ($n = 12$). The concept of participation is seen as relevant as both a means and as an end.

Participation as a means aims at more effective implementation of urban planning programmes and projects through active citizen involvement in project implementation, usually by means of labour and/or financial or in-kind contributions [42,60]. Participatory risk zone planning is an example in Japan highlighted by Aoki [60], whereby the reconstruction process was empowered through local community participation in a more effective manner.

Participation as an end implies that citizens come up with ideas, take part in the decision-making process, assume responsibility, and finally arrive at self-management. Baudoin [61] revealed the importance of a community centric approach, where the grass-roots level and vulnerable communities paved the way for the design and application of early warning systems in an urban environment.

Several studies emphasised the inclusion of different layers of the community in the planning process, especially vulnerable and marginalized groups, as well as different age groups, genders, and ethnicities [54,61]. Castro [51] further discussed this with a specific approach to work with population with disabilities, noting the importance of incorporating them in all disaster risk and emergency management processes.

Institutional collaboration is also highlighted as a major concern for planning and implementation of urban planning strategies. For example, government should empower the institutions and facilitate them in the process [62]. Two studies highlighted the importance of a common platform, where experts and their institutions could present their plans, findings, and innovations in urban planning [63,64]. As highlighted by Djalante [63], effective participation and meaningful collaboration among all stakeholders, governments, NGOs, and communities is an important element of disaster risk reduction, where the increased recognition of the roles and responsibilities of local stakeholders in managing disaster risks is crucial. Similarly, the media is also recognised as a stakeholder that can, for example, be used to empower the early warning, awareness, and hazard education systems, which can each play an important role in urban planning for tsunami-prone regions [39].

3.2.3. Resilience Thinking Approaches

Numerous studies have suggested resilience thinking to support emergence of the resilient city concept, such as through the development of appropriate capacities of the urban system based on a combination of "absorbing disturbances and achieving a balance". An example of this is nature-based solutions ($n = 5$). The green city concept by Arif [57], the Waju community model by Ueda [65], and the build back greener approach by Mabon [66], all highlighted the importance of nature-based solutions in urban disaster resilience as a way to reduce disaster risk in tsunami-prone areas. As suggested by Arif [57], the urban development process should be conducted in a planned and integrated way, with special attention to spatial and environmental aspects to ensure an efficient urban management. The approach was further discussed with an ecological perspective by Ahern [67], with five strategies to build resilience capacity and transdisciplinary collaboration: biodiversity, urban ecological networks and connectivity, multifunctionality, redundancy and modularization, and adaptive design.

Community participation was also included in three studies as a key component in developing resilience. For example, Kennedy et al. [68] proposed a resilience thinking approach, as well as community and stakeholder involvement. They suggested the phrase "build back safer" instead of the "build back better", as they argued that the idea of "better" had multiple interpretations.

Ueda [66], in their study of the "Waju" community in Japan, stressed the importance of traditional knowledge, which must be combined with modern advanced technologies. They suggested that this traditional knowledge could be used to promote autonomy for disaster prevention and local supply of infrastructure and logistics for adaptation to hazardous environments.

Most of the reviewed sources stressed a "thinking ahead approach", integrating relief and development through long-term disaster risk reduction with sustainable urban development processes. Similarly, the knowledge required for approaching this urban sustainability and resilience can evolve rapidly in an adaptive planning and design context, as a complement to urbanization processes and projects in coastal regions.

3.2.4. Hard and Soft Engineering

Hard engineering management involves using artificial structures, whereas soft engineering management is a more sustainable and natural approach to manage coastal areas, such as to tackle coastal erosion [69]. The literature identified a range of potential countermeasures for disaster risk reduction in these broad categories: (1) hard engineering, containing seawall, groynes, and other hard engineered structures [70]; (2) soft engineering, including beach nourishment, mangrove afforestation [57,71,72], coral reef transplant [73], and coastal forest plantation [71]; (3) combined measures, such as beach nourishment seawall/groynes/breakwaters [74].

As per the reviewed sources, seawalls and manmade barriers such as groynes have been frequently used in coastal regions ($n = 3$). Nateghi [71] further explained that large seawalls have been shown to have been effective at reducing both mortality and damage rates in a hazard event, but smaller seawalls (around 5 m high) showed no effectiveness in reducing impact. However, there were a lack of studies on utilizing other concrete or wooden structures under hard engineering in these regions. However, as highlighted in some studies, soft engineering and/or combined structures are increasingly popular worldwide. For example, these can involve beach nourishment and biological restoration [73]. Among them, mangroves and coral reefs were prominent features in the reviewed sources ($n = 5$). Takagi [73] concluded that the impact of coastal floods could be substantially mitigated by planting a mangrove belt in front of a dyke. Soft engineering structures can also allow for sediment movement along the shoreline and often try to replicate natural processes. Some of the studies revealed the utilization of nature-based solutions with soft engineering structures as a proven sustainable solution for urban planning in tsunami-prone areas [73]. These authors contend that nature-based solutions have a prominent role in building resilience to future tsunamis, and can simultaneously act as a site for education and memorialisation. Further, coastal forests, coral reefs, and compacted dunes and wet lands in these urban areas could be incorporated as natural buffers for hazard mitigation, which could provide long-term sustainable solutions to face these extreme scenarios [54].

3.3. Urban Planning in Tsunami-Prone Areas of Sri Lanka

This section considers the findings in relation to RQ2 and RQ3 from the empirical investigations in Sri Lanka. Sri Lanka suffered the second highest losses in absolute terms from the 2004 Indian Ocean Tsunami. Coastal cities of the country were the most affected areas during the event including Ampara, Galle, Matara, Colombo, and Hambanthota [75]. The need for better urban planning to address disaster risk in the coastal cities of Sri Lanka was stressed in several reviews following the disaster. The section begins with an overall reflection of the urban planning strategies being adopted in Sri Lanka. The subsequent subsections discuss in more detail each of the main urban planning strategies, as well as gaps and limitations in the strategies which are identified through the expert interviews and focus group discussion.

Similar to the international literature, respondents pointed to evacuation planning and spatial planning as the major urban planning strategies in Sri Lanka to address disaster risk in coastal regions. For example, evacuation planning has been empowered with community awareness programmes, especially after the 2004 Indian Ocean Tsunami. This includes periodic drills in tsunami-prone areas since 2005. However, several disaster management professionals cautioned that such awareness programmes and drills had been badly disrupted or discontinued due to COVID-19.

The country has also implemented buffer zone regulations and guidelines as a spatial planning strategy for tsunami preparedness. However, as revealed by several of the civil engineering and disaster management professionals, under a new set of rules envisaged by the government, the buffer will be reduced to between 55 and 25 m in the southern districts and from 100 to 50 m in the northeast. Seven respondents also commented on the weak implementation of these regulations in coastal regions in Sri Lanka. For example, a civil engineer noted, "If there are regulations, planners have to adhere. But when it comes to guidelines, some can follow and some can skip" (C1). Further, an architecture professional stated that, "because of the political influence, they misinterpret and get away without following them".

The interviews also revealed that most of the planning strategies address a single hazard event, but do not adequately address multi-hazard scenarios. All twelve experts highlighted that law making and implementing authorities had tended to focus on single hazard events, with many attributing this to the different hazard profiles across the country, and many regions facing a dominant hazard that had become a focus for their preparedness efforts. Nine of the respondents agreed that a multi-hazard approach is important in planning. Several used the example of COVID-19 to illustrate why a multi-hazard approach is needed, as the country had experienced flooding and coastal storms during the pandemic. Several noted that Sri Lankan legal frameworks did not adequately address pandemic preparedness and multi-hazard scenarios into planning.

Many of the respondents recognised a need to combine hard and soft engineering strategies in coastal regions as a way to reduce disaster risk associated with hazards such as tsunamis, floods, and coastal storms. However, several town planning and architecture-related professionals in particular stressed that, currently, there is an underutilisation of soft engineering and nature-based solutions in tsunami-prone areas. A disaster management professional explained (D2), "People look for short term solutions rather than the long-term benefits in planning". This was further explored in the focus group discussion, which revealed that land scarcity in coastal regions and the mindset of people and the government are the main reasons behind weak adoption of these approaches.

3.3.1. Spatial Planning in Sri Lanka

The importance of spatial planning in tsunami-prone areas has been recognised in Sri Lanka, and land use regulations have been introduced in such regions. Guidelines for housing development in coastal areas of Sri Lanka highlighted statutory requirements and best practice guides for settlement planning, housing design, and service provision for hazard preparedness in general. As observed in many international studies, digital technologies have also frequently been used in Sri Lanka for hazard and vulnerability mapping. The maps have been developed by the National Building and Research Organization and similar national level organizations. For tsunamis specifically, shoreline analysis, risk assessments, and vulnerability mapping exercises were also performed in coastal regions for disaster risk reduction after the Indian Ocean Tsunami. Despite such maps being developed, ten of the respondents highlighted that the maps are typically not used at the local authority level. The focus group discussion further highlighted that this was likely due to a combination of poor understanding and knowledge among local officials, but also insufficient effort to provide access to such maps for officials or the wider public. Further concerns were expressed about the resources available to update such maps, noting that many were developed over 15 years ago, often with the support of international agencies after the 2004 tsunami.

Another widely expressed concern was the weak implementation of building and planning regulations by developers, and a lack of monitoring by the local authorities. A town planning specialist (T2) summarised this as, "Planning and plan implementation are different" and "enforcers are abusing the power". Further, an architect and town planner (A2) suggested the absence of clear policy can give builders the basis to ignore the guidelines.

However, some also highlighted resource constraints as a major barrier to implementation. For example, several noted that land scarcity, cost, and a lack of expertise among local authorities as the barriers for proper land suitability analysis and spatial planning prior to construction. An experienced environmentalist and town planner (T1) who had worked in local authorities elaborated, "For smaller buildings, soil testing is a great cost. For smaller lands, the recommended shape of the building cannot be followed. Land scarcity and financial constraints are the barriers for proper implementation".

3.3.2. Community and Stakeholder Participation in Sri Lanka

In common with the wider literature, all respondents agreed that stakeholder participation is very important in urban planning in order to effectively address disaster risk. However, many doubted the effectiveness of current approaches in Sri Lanka. Public consultations were identified by many respondents as the prevailing platform for the public to comment on development proposals in the country, and this was recognised as a fundamental step in participatory risk zone planning. An architecture professional (A2) noted that, "Public consultation is happening, but mostly through conventional interest group meetings". In order to increase the reach of such consultations, during the focus group it was suggested that digital platforms, including social media, could be better utilised in this process.

Poor access to risk information was also highlighted as a barrier to participation. Respondents tended to concur that the general public had no easy access to hazard, vulnerability, or risk information. Some respondents suggested that this might limit the ability for the public to make informed contributions to the planning process as they might not have access to the evidence that could be used to support their concerns.

Several respondents also highlighted the inadequate consideration of vulnerable communities in the planning process, for example, a failure to engage marginalised groups, people living in informal settlements, and people with disabilities. During the focus group, it was suggested that these people, as well as communities more widely, would benefit from greater engagement during the preplanning phases of development.

Some respondents warned that public consultation was currently hindered by political influence in the regions, although they did not feel comfortable elaborating on their concerns.

Greater involvement and accountability of related industries, fisheries, and tourism in the planning process was recommended by many experts. For example, one green building specialist and town planner (G1) suggested that, "Industries should be provided with and subject to enforcement against clearer regulations. This would force them to be more educated and prepared to self-analyze the impacts of their activities along the coastline".

Reflecting upon many of these concerns, during the focus group there was recognition of a need to re-engineer the stakeholder engagement processes, and provide better platforms to facilitate such engagement.

3.3.3. Resilience Thinking Approaches in Sri Lanka

Many cities in Sri Lanka have joined the UN Making Cities Resilient campaign, while similar to many other countries that have experienced major disasters in their recent history, Sri Lanka has also implemented a "build back better" approach in development, especially after the 2004 tsunami. More recently, this has been refined to be a "build back safer" approach to promote the long-term protection of the community. Although related terms were frequently used during the interviews and focus group, several respondents were sceptical and indicated that only a few major projects, such as the Galle City Development Project, properly adhered to these principles. They noted that many local authority level projects, especially those without international donor support, did not adopt "build back better" or other resilience-based approaches at the city or project levels.

The "living with floods" concept is visible in a few rural areas, where native solutions and indigenous knowledge have been integrated within survival planning. Despite some isolated examples, a majority of respondents felt more effort should be made to draw upon lessons from the past, especially drawing upon local knowledge. Several noted that current planning approaches are top-down, and fail to adequately draw upon local experiences when developing resilience. The importance of combining scientific and traditional knowledge was also highlighted by some. For example, an Architect (H1) suggested, "Cases can be found in rural coastal cities where tidal patterns, flood information, and mitigation measures have been adopted from local knowledge. Boat evacuation during floods also shows the validity of traditional methods even in the present context". Other resilience examples cited by the experts included the ancient cascading system, ecological restoration. and community resilience with social capital that could be effective in an urban planning context.

In order to promote greater resilience thinking in urban planning, several respondents stressed the importance of leadership. This was especially so for major city development projects and national level planning, which invariably require the support of senior government officials. A disaster management specialist (D1) proposed that, "Policy makers, politicians, and planners should get together and do proper planning after proper analysis. Good leaders will enable good resilience thinking approaches".

3.3.4. Hard and Soft Engineering Approaches in Sri Lanka

Sri Lanka has adopted a range of hard and soft engineering strategies to protect its coastal regions. Groynes and sea walls are among the major hard engineering approaches that have been used, along with soft strategies such as beach nourishments and mangrove plantations. However, most experts noted the absence of many large hard engineered structures.

One structural engineer (C2) explained, "Sri Lanka doesn't need many coastal protection structures—there are cost constraints and tsunami is such a low frequency event. It would be difficult to justify. Instead, hazard resilient building construction should be given priority".

The importance of having an integrated building code was stressed by all the structural engineering respondents. However, one (C3) highlighted some limitations in the current approach, including a lack of consistency and legal obligations, "Several guidelines and frameworks from different authorities which are not legally binding will not serve the purpose".

Recently, nature-based solutions have been introduced as a more sustainable solution by the respective agencies. Some examples of ecological restoration programmes in coastal regions were highlighted by respondents as evidence of the emerging shift towards nature-based solutions, and of further opportunities to make better use of existing natural resources to support safer urban planning in tsunami-prone areas. These approaches appear to be popular with authorities as they are often viewed as cost effective when compared with hard engineering. Despite this, one respondent cautioned that there are often barriers, and gave the example of a protection of plantation scheme that was at risk due to land ownership issues and a lack of understanding among communities. Competing stakeholder interests were also seen as a barrier. For example, five respondents stressed the need for more stringent legal frameworks to protect natural assets that were providing protection, including beaches and coral reefs. One expert suggested that, "There must be dedicated personnel at local authorities with the authority to monitor the preservation of beach and nature-based solutions, and report to the authorities to take actions against those who violate the laws".

4. Discussion

In order to bring the literature and empirical findings together and address RQ2, Figure 8 illustrates some of the key urban planning strategies that are currently used in tsunami-prone areas before a hazard event occurs, in the immediate response to a hazard event, and during recovery and reconstruction following a disaster. The specific urban planning approaches are broadly grouped into six high-level strategies: community participation, spatial planning, soft and hard engineering, evacuation planning, and resilience thinking. Some of these strategies are specific to one phase, while other strategies cross multiple phases, although the specific urban planning approaches within each strategy tend to differ across them. Within each strategy, examples of the types of approaches are presented.

RQ3 sought to identify gaps or limitations in these existing approaches. The following discussion attempts to draw out some common themes to emerge.

The importance of community participation in planning processes is widely recognized in the literature across all three phases, i.e., before, responding, and recovering from disaster. Despite also recognizing its importance, existing approaches in Sri Lanka have often failed to reach members of the community, especially marginalized and vulnerable groups. The methods of engagement are limited in nature and there has been little attempt to increase the reach, such as through new technology or social media. Most of the reviewed sources have also highlighted the importance of stakeholder engagement in a broader sense and to provide a realistic picture in grassroots level disaster risk reduction. The integration of different stakeholders' perspectives is a particular concern in the planning process, and many studies, including Sri Lanka, have highlighted the potential for conflict due to different interests.

Similarly, a lot of disaster risk information is gathered in Sri Lanka, but is difficult to access, and often not used to inform development planning. It is also evident that although there was some investment in technology and the development of risk information following the 2004 tsunami, there has been only limited use of GIS and remote sensing, or resources available to adopt more recent developments such as web-based planning and better use of real-time information. This reaffirms the challenges such countries face in developing sustainable approaches to urban planning. Although they can benefit from substantial external assistance following a major disaster, the country is often left without the capacity to sustain these efforts and, over time, the quality and timeliness of such information diminishes.

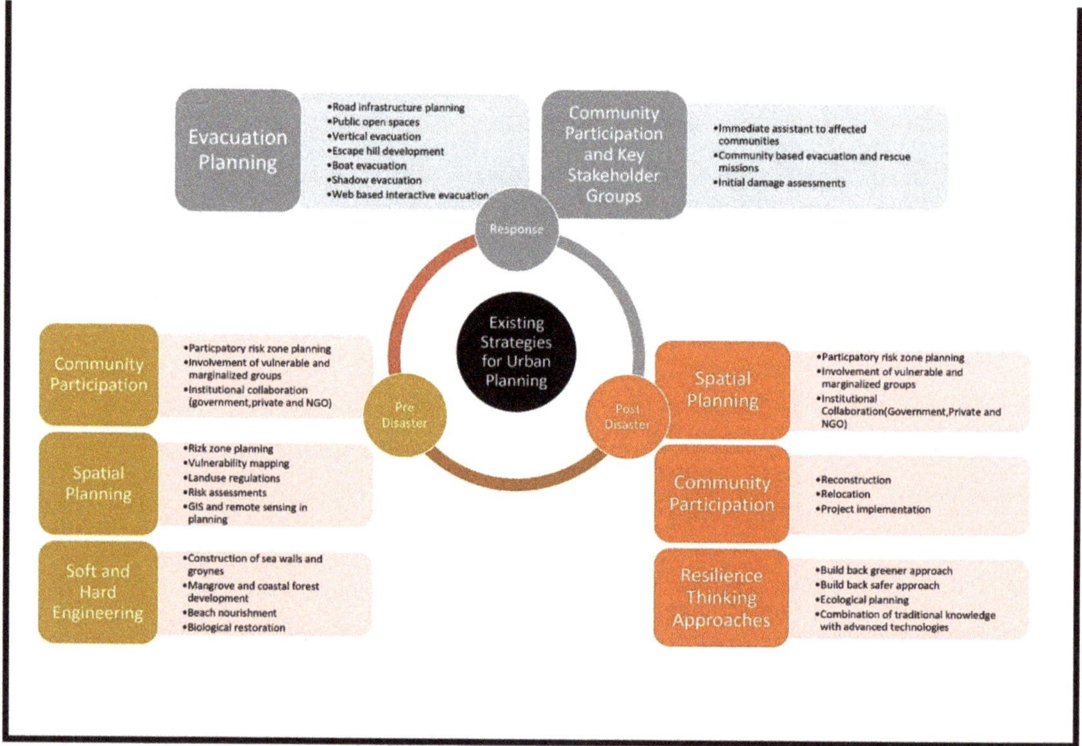

Figure 8. Existing strategies for urban planning in tsunami-prone areas.

Among the literature, the lack of health- and population-related planning strategies in the predisaster phase is notable, for example, locational and population growth planning. This is also the case in Sri Lanka. For example, the role of population planning aspects in highly populated coastal cities including Colombo, Kalutara and Galle, does not appear to have been actively considered in relation to disaster risk.

A combination of soft and hard engineering approaches can be effective in reducing disaster risk in tsunami-prone areas. The potential for hard engineering structures is well established, although in many situations, the cost is prohibitive, especially when facing very infrequent hazard events, which makes it difficult to justify. Sri Lanka appears to have embraced some soft engineering, including nature-based solutions that are increasingly being promoted for their potential benefits in biodiversity protection, safety, and cost effectiveness. However, there are concerns about the extent to which such interventions are being maintained and the capacity of local officials to protect them.

Even though resilience thinking approaches were discussed in the literature, most of the existing approaches were limited to ecological resilience or build back better approaches after a disaster. Social and physical aspects were not included in most of the studies [73]. To facilitate a holistic interpretation of urban planning in these coastal regions, it may be necessary to employ different aspects of resilience in the existing resilience approaches and methods as well. For instance, assessing ecosystem services of a neighborhood's green infrastructure may reveal the performance of a particular urban area in relation to biodiversity protection, public open space utilization, and local level engagement. This supports the view of Ueda [65] who suggested that it would be important to open a space for DRR practitioners and scholars to pause for thought, to reconsider, and to reformulate. This might include providing opportunities for knowledge sharing and coupling of traditional and scientific knowledge in resilience thinking for better results.

The need for a more multi-hazard approach is widely recognized in the Sendai Framework for Disaster Risk Reduction. However, a commonality among previous studies from the literature and current practices in Sri Lanka is the primary focus on single hazard events, or at best, some consideration of cascading hazards, where one hazard, for example, an earthquake, triggers another hazard, a tsunami. The COVID-19 pandemic has again brought to the forefront the importance and challenges associated with multi-hazard threats. These can be especially challenging where they are concurrent but independently occurring hazards that interact. The nature of COVID-19, which has disrupted society for several years, has clearly exposed the potential for this type of situation. Many countries around the world have experienced other hazard events while tackling the pandemic threat. As highlighted in Sri Lanka, the COVID-19 response resulted in a shift of priorities, alterations in work practices and locations, the imposition of physical distancing, self-isolation and quarantine measures, as well as temporary lockdowns of entire communities. It will be important to understand how such changes might impact wider hazard vulnerability, and to what extent urban planning can reduce or exacerbate disaster risk in such a situation.

This also relates to the broader issue of systemic risk, which is highlighted in the 2022 Global Assessment Report [76]. The concept of systemic risk is based on the notion that the risk of an adverse outcome of a policy, action, or hazard event can depend on how the elements of the affected systems interact with each other. This can either aggravate or reduce the overall effect of the constituent parts. Interactions occur through positive or negative feedback processes. Systemic risk creates the chance of system malfunction or even collapse [77]. Urban planning has the potential to greatly influence risk-sensitive urban development in a manner that can transform the way cities are built to face the uncertainties that arise from climate-induced disaster risks, but also address the complexity of multi-hazard threats. However, there appears to be a dearth of previous studies that attempt to tackle such complexity, and there is little evidence in Sri Lanka that such approaches are permeating into policy and practice.

5. Conclusions

This systematic review of previous studies on urban planning in tsunami-prone areas, along with a more detailed examination of current urban planning approaches in Sri Lanka, provide some useful insights into the range of strategies available to reduce disaster risk, but also some of the many challenges associated with their implementation in policy and practice.

Tsunamis, as an infrequent, but potentially devastating hazard threat to many coastal cities, have prompted researchers and related policymakers to work on urban planning strategies that can reduce disaster risk in tsunami-prone areas. It is also evident from the literature and results to emerge from the study in Sri Lanka that the 2004 Indian Ocean tsunami, as well as other major tsunami events in recent years, have prompted increased investment and attention on urban planning efforts, as well as numerous studies to investigate specific approaches and their effectiveness. However, the lack of common methodological approaches across studies, as well as the lack of detail in the presentation of research methods, makes direct comparisons across studies more difficult. It also does not lend itself to research synthesis, which could otherwise be used to inform improvements to urban planning in different contexts.

The results also reveal there are many urban planning challenges to overcome in order to better protect tsunami-prone areas. The model in Figure 8 provides a useful way of illustrating the breadth of urban planning strategies and approaches that can be adopted to help tackle disaster risk in tsunami-prone areas. The model is also an initial attempt to provide a basis for further detailed empirical studies of urban planning in tsunami-prone areas, including much needed comparative studies. However, the authors recognize there is much work to be done in further exploring the model through more detailed empirical studies in different contexts and to ensure that current practices are adequately reflected.

This review is limited by its reliance on other scholars self-reporting results in their own studies and opinions from the different professionals in Sri Lanka. Furthermore, there are language inconsistencies across the literature in terms of what scholars identify as soft engineering, nature-based, and hard engineering. Therefore, the analysis of the literature is based on our interpretation of what is at times unclear work of others. Still, this review points to new research that would benefit the field of urban planning in tsunami-prone areas. Primarily, future research will need to establish guiding planning principles, build goal-oriented asassessment frameworks under these principles, and test the frameworks with empirical assessment studies. This should be pursued in parallel with the development of a common framework to unify urban planning literature. A common set of urban planning strategies can emerge for multi-hazard scenarios and to tackle systemic risk.

Author Contributions: Conceptualization, U.T.G.P., R.H. and D.A.; methodology, R.H. and D.A.; software U.T.G.P.; validation, C.D.Z. and A.A.S.E.A.; formal analysis, U.T.G.P.; investigation, U.T.G.P.; resources, U.T.G.P., C.D.Z. and A.A.S.E.A.; data curation, U.T.G.P. and C.D.Z.; writing—original draft preparation, U.T.G.P.; writing—review and editing, R.H.; visualization, U.T.G.P.; supervision, R.H., D.A. and R.D.; project administration, R.H., D.A. and R.D.; funding acquisition, R.H. and D.A. All authors have read and agreed to the published version of the manuscript.

Funding: The underpinning research was co-funded by UK Research and Innovation (UKRI) through the UK Government's Global Challenges Research Fund (GCRF) and the UK Newton Fund [grant number EP/V026038/1]. The underpinning research was co-funded by the UK Parliamentary Under Secretary of State for Business, Energy and Industrial Strategy (BEIS) through the Newton Prize and UK Newton Fund [grant number AM002715]. UKRI and BEIS accept no liability, financial or otherwise, for expenditure or liability arising from the research funded by the grant, except as set out in the Terms and Conditions or otherwise agreed in writing.

Institutional Review Board Statement: The research was approved by the ethics committee of the lead University that coordinated the research. Institution: University of Huddersfield School of Applied Sciences Research Integrity and Ethics Committee Code: SAS-SREIC 25.01.21-9.

Informed Consent Statement: Informed consent to participate in the study was obtained from participants in interviews and the focus group. This included information on how data would be protected, confidentiality and privacy, as well consent to publish results in an anonymized form. No personally identifiable information is included within this publication.

Data Availability Statement: The data that support the findings of this study are available on request from the corresponding author, U.T.G.P. The data are not publicly available due to their containing information that could compromise the privacy of research participants.

Acknowledgments: The authors gratefully acknowledge the contributions of experts in interviews and the focus group discussion.

Conflicts of Interest: The authors declare no conflict of interest.

Appendix A

Table A1. Reviewed sources of the systematic review.

ID No.	Paper	Title of the Journal and Published Source	Author	Year
1	Quantifying urban physical growth types in Banda Aceh City after the 2004 Indian Ocean Tsunami	E3S Web of Conferences	Amri, S.R. Giyarsih	2020
2	Green city Banda Aceh: City planning approach and environmental aspects	IOP Conference Series: Earth and Environmental Science	A.A. Arif	2017
3	Escape hill development as a strategy to improve urban safety after earthquake and tsunami Aceh 2004 based on regional planning and geotechnical aspect	*Journal of Physics*: Conference Series	M. Munirwansyah, H. Munirwan, M. Irsyam, R.P. Munirwan	2020
4	Resident's satisfaction to relocated Houses after 2004 Indian Ocean Tsunami, Thailand	*Procedia Engineering*, Elsevier	Titaya Sararita, Kondo Tamiyob, and Elizabeth Maly	2017
5	Integration of disaster management strategies with planning and designing public open spaces	*Procedia Engineering*, Elsevier	R.R.J.C. Jayakody, D. Amarathunga, R. Haigh	2017
6	Assessment of road traffic performance of the Tsunami evacuation road in Padang Municipality area based on the traffic volume simulation approach	E3S Web of Conferences	Y. Yossyafra, N. Fitri, R.P. Sidhi, Y. Yosritzal, D.I. Mazni	2020
7	Opportunities and Risks of the "New Urban Governance" in India: To What Extent Can It Help Addressing Pressing Environmental Problems?	*Journal of Environment & Development*, SAGE	Jeroen van der Heijden	2016
8	A Study on transformation of living environment and domestic Spatial Arrangements: Focused on a western coastal housing settlement of Sri Lanka after Sumatra Earthquake and Tsunami 2004	*Journal of Asian Architecture and Building Engineering*, Architectural Institute of Japan	Woharika Kaumudi Weerasinghe and Tsutomu Shigemura	2008

Table A1. *Cont.*

ID No.	Paper	Title of the Journal and Published Source	Author	Year
9	Urban landscape sustainability and resilience: The promise and challenges of integrating ecology with urban planning and design ****	*Landscape Ecology*, Springer	Jack Ahern	2013
10	The Sendai Framework for Disaster Risk Reduction: Renewing the global commitment to people's resilience, health, and well-being	*International Journal of Disaster Risk Science*, Springer	Amina Aitsi-Selmi, Shinichi Egawa, Hiroyuki Sasaki, Chadia Wannous, Virginia Murray	2015
11	"Waju" and its evolution with urban technology-Japanese sustainable community for disaster resilience	IOP Conference Series: Earth and Environmental Science	R. Ueda	2019
12	Enhancing post-disaster resilience by 'building back greener': Evaluating the contribution of nature-based solutions to recovery planning in Futaba County, Fukushima Prefecture, Japan	*Landscape and Urban Planning*	L. Mabon	2019
13	From top-down to "community-centric" approaches to early warning systems: Exploring pathways to improve disaster risk reduction through community participation	*International Journal of Disaster Risk Science*, Springer	Marie-Ange Baudoin, Sarah Henly-Shepard, Nishara Fernando, Asha Sitati Zinta Zommers	2016
14	Disaster risk perception in urban contexts and for people with disabilities: case study on the city of Iquique (Chile)	*Natural Hazards*, Springer	Carmen-Paz Castro, Juan-Pablo Sarmiento, Rosita Edwards, Gabriela Hoberman, Katherine Wyndham	2016
15	Bangkok to Sendai and Beyond: Implications for disaster risk reduction in Asia	*International Journal of Disaster Risk Science*, Springer	Ranit Chatterjee, Koichi Shiwaku, Rajarshi Das Gupta, Genta Nakano, Rajib Shaw	2015
16	The Indian Ocean Tsunami: Economic impact, disaster management, and lessons	Asian Economic Papers-The Earth Institute at Columbia University and the Massachusetts Institute of Technology	Prema-chandra Athukorala	2006
17	The COVID-19 pandemic: Impacts on cities and major lessons for urban planning, design, and management	*Science of the Total Environment*, Elsevier	Ayyoob Sharifi, Amir Reza, Khavarian-Garmsir	2020
18	Flood risk and adaptation in Indian coastal cities: Recent scenarios	*Applied Water Science*, Springer	Ravinder Dhiman, Renjith VishnuRadhan, Eldho, Arun Inamdar	2018
19	Community resilience and urban planning in tsunami-prone settlements in Chile	*Disasters*, Overseas Development Institute	Marie Geraldine Herrmann-Lunecke	2019

Table A1. *Cont.*

ID No.	Paper	Title of the Journal and Published Source	Author	Year
20	Adaptive governance and managing resilience to natural hazards	*International Journal of Disaster Risk Science*, Springer	Riyanti Djalante, Cameron Holley, and Frank Thomalla	2019
21	Multi-criteria location planning for public facilities in tsunami-prone coastal areas	*OR Spectrum*, Springer	Karl F. Doerner, Walter J. Gutjahr, Pamela C. Nolz	2008
22	Validating a tsunami vulnerability assessment model (the PTVA model) using field data from the 2004 Indian Ocean Tsunami	*Natural Hazards*, Springer	Dale Dominey-Howes and Maria Papathoma	2006
23	From multi-risk evaluation to resilience planning: The case of central Chilean coastal cities	*Water* (Switzerland)	P. Barría, M.L. Cruzat, R. Cienfuegos, J. Gironás, C. Escauriaza, C. Bonilla, R. Moris, C. Ledezma, M. Guerra, R. Rodríguez, A. Torres	2019
24	Evaluation of the reconstruction plans for tsunami victims in Malaysia	*Journal of Asian Architecture and Building Engineering*, Architectural Institute of Japan	F.S. Ling	2006
25	The role of built environment's physical urban form in supporting rapid tsunami evacuations: Using computer-based models and real-world data as examination tools	*Frontiers in Built Environment*, Frontiers Editorial Office	Foong Sau Ling, Yoshimitsu Shiozaki, and Yumiko Horita	2018
26	Improved coastal zone planning and management	Integrated coastal zone planning in Asian tsunami-affected countries	Robert Kay	2006
27	Personal sky equipment for inhabitants of coastal cities: Envisioning an evacuation system to reduce disaster's impact during the climate change era	IOP Conference Series: Materials Science and Engineering	K. Januszkiewicz	2019
28	Urban resources selection and allocation for emergency shelters: In a multi-hazard environment	*International Journal of Environmental Research and Public Health*, Molecular Diversity Preservation Internationa (MDPI)	Wei Chen, Guofang Zhai, Chongqiang Ren, Yijun Shi, and Jianxin Zhang	2018
29	Challenges of post-tsunami reconstruction in Sri Lanka: Health care aid and the Health Alliance	CRISIS—FOR DEBATE, *The Medical Journal of Australia* (MJA)	Paul A Komesaroff and Suresh Sundram	2006
30	Assessing the impact of the Indian Ocean Tsunami on households: A modified domestic assets index approach	*Disasters*, Overseas Development Institute	Sudha Arlikatti, Walter Gillis Peacock, Carla S. Prater, Himanshu Grover, and Arul S. Gnana Sekar	2010

Table A1. *Cont.*

ID No.	Paper	Title of the Journal and Published Source	Author	Year
31	Sustainable downtown development for the tsunami-prepared urban revitalization of regional coastal cities	*Sustainability* (Switzerland)	T. Ito, T. Setoguchi, T. Miyauchi, A. Ishii, N. Watanabe	2019
32	Integrated approach for coastal hazards and risks in Sri Lanka	*Natural Hazards and Earth System Sciences*, Copernicus Publications on behalf of the European Geosciences Union	M. Garcin, J.F. Desprats, M. Fontaine, R. Pedreros, N. Attanayake, S. Fernando, C.H.E.R. Siriwardana, U. De Silva, and B. Poisson	2008
33	Tsunami hazard assessment of Chabahar Bay related to megathrust seismogenic potential of the Makran subduction zone	*Natural Hazards*, Springer	A.R. Payande, M.H. Niksokhan, H. Naserian	2014
34	Risk and vulnerability assessment to tsunami and coastal hazards in Indonesia: Conceptual framework and indicator development	Conference Paper, Research Gate	J. Post, K. Zosseder, G. Strunz, J. Birkmann, N. Gebert, N. Setiadi, H.Z. Anwar, H. Harjono, M. Nur, T. Siagian	2014
35	An integrated social response to disasters: The case of the Indian Ocean Tsunami in Sri Lanka	*Disaster Prevention and Management*, Emerald Group Publishing Limited	Siri Hettige and Richard Haigh	2016
36	Assessing people's early warning response capability to inform urban planning interventions to reduce vulnerability to tsunamis	*Institute of Geodäsey and Geoinformation*, Bonn University	N.J. Setiadi	2014
37	Tsunami vulnerability assessment in urban areas using numerical model and GIS	*Natural Hazards*, Springer	Tune Usha, M.V. Ramana Murthy, N.T. Reddy, Pravakar Mishra	2011
38	Measuring tsunami planning capacity on US Pacific coast	*Natural Hazards Review*, Infrastructure Resilience Division	Z. Tang, M.K. Lindell, C.S. Prater, and S.D. Brody, 2008.	2008
39	Sequencing and combining participation in urban planning: The case of tsunami-ravaged Onagawa Town, Japan	*Cities*, Elsevier	N. Aoki	2018
40	Urban planning and tsunami impact mitigation in Chile after February 27, 2010	*Natural Hazards*, Springer	M.G.H. Lunecke	2015
41	The meaning of 'build back better': Evidence from post-tsunami Aceh and Sri Lanka	*Journal of Contingencies and Crisis Management*	J. Kennedy, J. Ashmore, E. Babister, and I. Kelman	2008
42	Environmental implications for disaster preparedness: Lessons Learnt from the Indian Ocean Tsunami	*Journal of Environment Management*, Elsevier	H. Srinivas	2008

Table A1. *Cont.*

ID No.	Paper	Title of the Journal and Published Source	Author	Year
43	A systematic study of disaster risk in brunei darussalam and options for vulnerability-based disaster risk reduction	*International Journal of Disaster Risk Science*, Springer	Anthony Banyouko Ndah, John Onu Odihi	2017
44	Disaster waste clean-up system performance subject to time-dependent disaster waste accumulation	*Natural Hazards*, Springer	Cheng Cheng, Lihai Zhang, Russell George Thompson, Greg Walkerden	2017
45	Identification and classification of urban micro-vulnerabilities in tsunami evacuation routes for the city of Iquique, Chile	*Natural Hazards and Earth System Sciences*	G. Álvarez, M. Quiroz, J. León, R. Cienfuegos	2018
46	Heritage planning and rethinking the meaning and values of designating heritage sites in a post-disaster context: The case of Aceh, Indonesia	IOP Conference Series: Earth and Environmental Science	Z.D. Meutia, R. Akbar, D. Zulkaidi	2018
47	Assessing tsunami vulnerability areas using satellite imagery and weighted cell-based analysis	*International Journal of GEOMATE*	Guntur, A.B. Sambah, F. Miura, Fuad, D.M. Arisandi	2017
48	Mangrove forest against dyke-break-induced tsunami on rapidly subsiding coasts	*Natural Hazards and Earth System Sciences*	H. Takagi, T. Mikami, D. Fujii, M. Esteban, S. Kurobe	2016
49	A household-level flood evacuation decision model in Quezon City, Philippines	*Natural Hazards*, Springer	Ma. Bernadeth, Lim, Hector R., Lim Jr, Mongkut Piantanakulchai, Francis Aldrine Uy	2016
50	Visual exploration of tsunami evacuation planning	*Journal of the Visualization Society of Japan*	Cui Xie, Guangxiao Ma, Qiong Li, Jinjin Xun, Junyu Dong	2015
51	The 2011 Tohoku Tsunami: Implications for natural disaster management in Japan	*Revista INVI*	Y.C.C. Gatica, M.B. Benítez	2015
52	Statistical analysis of the effectiveness of seawalls and coastal forests in mitigating tsunami impacts in Iwate and Miyagi Prefectures	*PLOS ONE*	Roshanak NateghiJeremy D. Bricker, Seth D. Guikema, Akane Bessho	2016
53	Natural hazards, vulnerability and structural resilience: Tsunamis and industrial tanks	*Geomatics, Natural Hazards and Risk*, Informa UK Limited	Ahmed Mebarki, Sandra Jerez, Gaetan Prodhomme, and Mathieu Reimeringer	2016
54	Influence of road network and population demand assumptions in evacuation modeling for distant tsunamis	*Natural Hazards*, Springer	Kevin D. Henry, Nathan J. Wood, Tim G. Frazier	2016

Table A1. *Cont.*

ID No.	Paper	Title of the Journal and Published Source	Author	Year
55	Disaster risk reduction including climate change adaptation over South Asia: Challenges and ways forward	*International Journal of Disaster Risk Science*, Springer	Rajesh K. Mall, Ravindra K. Srivastava, Tirthankar Banerjee, Om Prakash Mishra, Diva Bhatt, Geetika Sonkar	2019
56	Tsunami risk reduction for densely populated Southeast Asian cities: Analysis of vehicular and pedestrian evacuation for the city of Padang, Indonesia, and assessment of interventions	*Natural Hazards*	M. Di Mauro, K. Megawati, V. Cedillos, B. Tucker	2013

References

1. Amaratunga, D.; Haigh, R.; Premalal, S.; Siriwardana, C.; Liyanaarachchige, C. *Report on Exercise Indian Ocean Wave 2020: An Indian Ocean-Wide Tsunami Warning and Communications Exercise*; UNESCO: Paris, France, 2020.
2. Haigh, R.; Sakalasuriya, M.M.; Amaratunga, D.; Basnayake, S.; Hettige, S.; Premalal, S.; Jayasinghe Arachchi, A. The upstream-downstream interface of Sri Lanka's tsunami early warning system. *Int. J. Disaster Resil. Built Environ.* **2020**, *11*, 219–240. [CrossRef]
3. Thomalla, F.; Larsen, R.K. Resilience in the context of tsunami early warning systems and community disaster preparedness in the Indian Ocean Region. *Environ. Hazards* **2010**, *9*, 249–265. [CrossRef]
4. United Nations. Department of Economic and Social Affairs. World Population Prospects. 2018; Available online: https://population.un.org/wup/ (accessed on 5 March 2022).
5. UNHCR. Practical Guidance for UNHCR Staff on IDP Protection in the Context of Disasters and the Adverse Effects of Climate Change. 2021, pp. 1–33. Available online: https://www.unhcr.org/ (accessed on 5 March 2022).
6. UNFCC. *Technologies for Averting, Minimizing and Addressing Loss and Damage in Coastal Zones*; United Nations: New York, NY, USA, 2020; Volume 1. Available online: https://unfccc.int/ttclear/misc_/StaticFiles/gnwoerk_static/2020_coastalzones/b9e88f6fea374d8aa5cb44115d201160/3863c9fabdf74ea49710189acbf6907a.pdf (accessed on 10 November 2021).
7. Li, L.; Switzer, A.D.; Wang, Y.; Chan, C.H.; Qiu, Q.; Weiss, R. A modest 0.5-m rise in sea level will double the tsunami hazard in Macau. *Sci. Adv.* **2018**, *4*, 1–12. [CrossRef] [PubMed]
8. Mori, N.; Takahashi, T.; Yasuda, T.; Yanagisawa, H. Survey of 2011 Tohoku earthquake tsunami inundation and run-up. *Geophys. Res. Lett.* **2011**, *38*, 6–11. [CrossRef]
9. Kajitani, Y.; Chang, S.E.; Tatano, H. Economic Impacts of the 2011 Tohoku-oki earthquake and tsunami. *Earthq. Spectra* **2013**, *29* (Suppl. S1), 457–478. [CrossRef]
10. IOC. *Tsunami Glosary*; Technical; Intergovernmental Oceanographic Commission: Paris, France, 2019; p. 85.
11. UNESCO. *Lesson Learnt from the 2018 Tsunamis in Palu and Sunda Strait*; UNESCO: Paris, France, 2019.
12. Goda, K.; Mori, N.; Yasuda, T.; Prasetyo, A.; Muhammad, A.; Tsujio, D. Cascading Geological Hazards and Risks of the 2018 Sulawesi Indonesia Earthquake and Sensitivity Analysis of Tsunami Inundation Simulations. *Front. Earth Sci.* **2019**, *7*, 261. [CrossRef]
13. Muhari, A.; Imamura, F.; Arikawa, T.; Hakim, A.R.; Afriyanto, B. Solving the Puzzle of the September 2018 Palu, Indonesia, Tsunami Mystery: Clues from the Tsunami Waveform and the Initial Field Survey Data. *J. Disaster Res.* **2018**, *13*, 2–4. [CrossRef]
14. Nakashima, M.; Lavan, O.; Kurata, M.; Luo, Y. Earthquake engineering research needs in light of lessons learned from the 2011 Tohoku earthquake. *Earthq. Eng. Eng. Vib.* **2014**, *13*, 141–149. [CrossRef]
15. UNESCO—IOC. *Global Ocean Science Report 2020*; UNESCO: Paris, France, 2021. [CrossRef]
16. Rus, K.; Kilar, V.; Koren, D. Resilience assessment of complex urban systems to natural disasters: A new literature review. *Int. J. Disaster Risk Reduct.* **2018**, *31*, 311–330. [CrossRef]
17. Huxley, M.; Inch, A. Urban Planning. In *International Encyclopedia of Human Geography*; Elsevier: Amsterdam, The Netherlands, 2020; pp. 87–92. [CrossRef]

18. León, J.; March, A. An urban form response to disaster vulnerability: Improving tsunami evacuation in Iquique, Chile. *Environ. Plan. B Plan. Des.* **2016**, *43*, 826–847. [CrossRef]

19. UNISDR. Hyogo framework for action 2005–2015. *Encycl. Earth Sci. Ser.* **2005**, *2005*, 508–516. [CrossRef]

20. UNISDR (United Nations International Strategy for Disaster Reduction). Sendai Framework for Disaster Risk Reduction 2015–2030. 2015. Available online: http://www.wcdrr.org/uploads/Sendai_Framework_for_Disaster_Risk_Reduction_2015-2030.pdf (accessed on 10 October 2021).

21. United Nations. *The 2030 Agenda for Sustainable Development*; United Nations: New York, NY, USA, 2015. [CrossRef]

22. CEPAL. *Planning for Disaster Risk Reduction within the Framework of the 2030 Agenda for Sustainable Development*; CEPAL: Santiago, Chile, 2021; 61p.

23. Alan, M.; Jorge, L. *Urban Planning for Disaster Risk Reduction: Establishing 2nd Wave Criteria*; University of Melbourne: Melbourne, Australia, 2013; pp. 1–13.

24. Kato, S.; Huang, W. Land use management recommendations for reducing the risk of downstream flooding based on a land use change analysis and the concept of ecosystem-based disaster risk reduction. *J. Environ. Manag.* **2021**, *287*, 112341. [CrossRef]

25. Wu, M.; Wu, Z.; Ge, W.; Wang, H.; Shen, Y.; Jiang, M. Identification of sensitivity indicators of urban rainstorm flood disasters: A case study in China. *J. Hydrol.* **2021**, *599*, 126393. [CrossRef]

26. Herrmann Lunecke, M.G. Urban planning and tsunami impact mitigation in Chile after February 27, 2010. *Nat. Hazards* **2015**, *79*, 1591–1620. [CrossRef]

27. Eisenman, D.P.; Cordasco, K.M.; Asch, S.; Golden, J.F.; Glik, D. Disaster planning and risk communication with vulnerable communities: Lessons from Hurricane Katrina. *Am. J. Public Health* **2007**, *97* (Suppl. S1), 109–115. [CrossRef]

28. Bosher, L. Built-in resilience through disaster risk reduction: Operational issues. *Build. Res. Inf.* **2014**, *42*, 240–254. [CrossRef]

29. Ma, C.; Wu, X.; Li, B.; Hu, X. The susceptibility assessment of multi-hazard in the Pearl River Delta Economic Zone, China. *Nat. Hazards Earth Syst. Sci. Discuss.* **2018**. *preprint*. [CrossRef]

30. Sengezer, B.; Koç, E. A critical analysis of earthquakes and urban planning in Turkey. *Disasters* **2005**, *29*, 171–194. [CrossRef]

31. Bathrellos, G.D.; Skilodimou, H.D. Land Use Planning for Natural Hazards. *Land* **2019**, *8*, 128. [CrossRef]

32. UN-Habitat. Cities and Pandemics: Towards a More Just, Green and Healthy Future. 2021. Available online: https://unhabitat.org/sites/default/files/2021/03/cities_and_pandemics-towards_a_more_just_green_and_healthy_future_un-habitat_2021.pdf (accessed on 21 March 2022).

33. ECLAC-UNDRR. The Coronavirus Disease (COVID-19) Pandemic: An Opportunity for a Systematic Approach to Disaster Risk for the Caribbean. Halodoc.Com. 2021. Available online: https://www.halodoc.com/artikel/begini-kronologi-lengkap-virus-corona-masuk-indonesia (accessed on 21 March 2022).

34. Moher, D.; Liberati, A.; Tetzlaff, J.; Altman, D.G. Preferred Reporting Items for Systematic Reviews and Meta-Analyses: The PRISMA Statement. *PLoS Med.* **2009**, *6*, e1000097. [CrossRef]

35. Guest, G.; Namey, E.; McKenna, K. How Many Focus Groups Are Enough? *Building an Evidence Base for Nonprobability Sample Sizes. Field Methods* **2017**, *29*, 3–22. [CrossRef]

36. Murray, N. *Urban Disaster Risk Governance: A Systematic Review*; EPPI-Centre: London, UK, 2017.

37. Dhiman, R.; VishnuRadhan, R.; Eldho, T.I.; Inamdar, A. Flood risk and adaptation in Indian coastal cities: Recent scenarios. *Appl. Water Sci.* **2019**, *9*, 5. [CrossRef]

38. Lim, M.B.B.; Lim, H.R.; Piantanakulchai, M.; Uy, F.A. A household-level flood evacuation decision model in Quezon City, Philippines. *Nat. Hazards* **2016**, *80*, 1539–1561. [CrossRef]

39. Athukorala, P.; Resosudarmo, B.P. The Indian Ocean Tsunami: Economic Impact, Disaster Management, and Lessons. *Asian Econ. Pap.* **2006**, *4*, 1–39. [CrossRef]

40. Di Mauro, M.; Megawati, K.; Cedillos, V.; Tucker, B. Tsunami risk reduction for densely populated Southeast Asian cities: Analysis of vehicular and pedestrian evacuation for the city of Padang, Indonesia, and assessment of interventions. *Nat. Hazards* **2013**, *68*, 373–404. [CrossRef]

41. Lagos, M.; Gutiérrez, D. Simulación del tsunami de 1960 en un estuario del centro-sur de Chile. *Rev. Geogr. Norte Gd.* **2005**, *33*, 5–18.

42. Van der Heijden, J. Opportunities and Risks of the "New Urban Governance" in India: To What Extent Can It Help Addressing Pressing Environmental Problems? *J. Environ. Dev.* **2016**, *25*, 251–275. [CrossRef]

43. Barría, P.; Cruzat, M.L.; Cienfuegos, R.; Gironás, J.; Escauriaza, C.; Bonilla, C.; Moris, R.; Ledezma, C.; Guerra, M.; Rodríguez, R.; et al. From multi-risk evaluation to resilience planning: The case of central Chilean coastal cities. *Water* **2019**, *11*, 572. [CrossRef]

44. Doerner, K.F.; Gutjahr, W.J.; Nolz, P.C. Multi-criteria location planning for public facilities in tsunami-prone coastal areas. *OR Spectr.* **2009**, *31*, 651–678. [CrossRef]

45. Yossyafra, Y.; Fitri, N.; Sidhi, R.P.; Yosritzal, Y.; Mazni, D.I. Assessment of road traffic performance of the Tsunami evacuation road in Padang Municipality area based on the traffic volume simulation approach. *E3S Web Conf.* **2020**, *156*, 04008. [CrossRef]

46. Januszkiewicz, K. Personal Sky Equipment for Inhabitants of Coastal Cities: Envisioning an Evacuation System to Reduce Disaster's Impact during the Climate Change Era. *IOP Conf. Ser. Mater. Sci. Eng.* **2019**, *471*, 082017. [CrossRef]
47. Henry, K.D.; Wood, N.J.; Frazier, T.G. Influence of road network and population demand assumptions in evacuation modeling for distant tsunamis. *Nat. Hazards* **2017**, *85*, 1665–1687. [CrossRef]
48. Xie, C.; Ma, G.; Li, Q.; Xun, J.; Dong, J. Visual exploration of tsunami evacuation planning. *J. Vis.* **2016**, *19*, 475–487. [CrossRef]
49. Usha, T.; Murthy, M.V.R.; Reddy, N.T.; Mishra, P. Tsunami vulnerability assessment in urban areas using numerical model and GIS. *Nat. Hazards* **2012**, *60*, 135–147. [CrossRef]
50. Amri, I.; Giyarsih, S.R. Quantifying urban physical growth types in Banda Aceh City after the 2004 Indian Ocean Tsunami. *E3S Web Conf.* **2020**, *200*, 07003. [CrossRef]
51. Castro, C.P.; Sarmiento, J.P.; Edwards, R.; Hoberman, G.; Wyndham, K. Disaster risk perception in urban contexts and for people with disabilities: Case study on the city of Iquique (Chile). *Nat. Hazards* **2017**, *86*, 411–436. [CrossRef]
52. Sararit, T.; Tamiyo, K.; Maly, E. Resident's satisfaction to relocated Houses after 2004 Indian Ocean Tsunami, Thailand. *Procedia Eng.* **2018**, *212*, 637–642. [CrossRef]
53. Payande, A.R.; Niksokhan, M.H.; Naserian, H. Tsunami hazard assessment of Chabahar bay related to megathrust seismogenic potential of the Makran subduction zone. *Nat. Hazards* **2015**, *76*, 161–176. [CrossRef]
54. Herrmann-Lunecke, M.G.; Villagra, P. Community resilience and urban planning in tsunami-prone settlements in Chile. *Disasters* **2020**, *44*, 103–124. [CrossRef]
55. Jayakody, R.R.J.C.; Amarathunga, D.; Haigh, R. Integration of disaster management strategies with planning and designing public open spaces. *Procedia Eng.* **2018**, *212*, 954–961. [CrossRef]
56. Sharifi, A.; Khavarian-Garmsir, A.R. The COVID-19 pandemic: Impacts on cities and major lessons for urban planning, design, and management. *Sci. Total Environ.* **2020**, *749*, 142391. [CrossRef] [PubMed]
57. Arif, A.A. Green city Banda Aceh: City planning approach and environmental aspects. *IOP Conf. Ser. Earth Environ. Sci.* **2017**, *56*, 012004. [CrossRef]
58. Meutia, Z.D.; Akbar, R.; Zulkaidi, D. Heritage planning and rethinking the meaning and values of designating heritage sites in a post-disaster context: The case of Aceh, Indonesia. *IOP Conf. Ser. Earth Environ. Sci.* **2018**, *158*, 012039. [CrossRef]
59. Guntur; Sambah, A.B.; Miura, F.; Fuad; Arisandi, D.M. Assessing tsunami vulnerability areas using satellite imagery and weighted cell-based analysis. *Int. J. GEOMATE* **2017**, *12*, 115–122. [CrossRef]
60. Aoki, N. Sequencing and combining participation in urban planning: The case of tsunami-ravaged Onagawa Town, Japan. *Cities* **2018**, *72*, 226–236. [CrossRef]
61. Baudoin, M.A.; Henly-Shepard, S.; Fernando, N.; Sitati, A.; Zommers, Z. From Top-Down to "Community-Centric" Approaches to Early Warning Systems: Exploring Pathways to Improve Disaster Risk Reduction Through Community Participation. *Int. J. Disaster Risk Sci.* **2016**, *7*, 163–174. [CrossRef]
62. Tang, Z.; Brody, S.D.; Tang, Z.; Lindell, M.K.; Prater, C.S.; Brody, S.D. Community and Regional Planning Program: Faculty Scholarly and Creative Activity Measuring Tsunami Planning Capacity on U.S. *Pacific Coast. Nat. Hazards Rev.* **2008**, *9*, 91–100. [CrossRef]
63. Djalante, R.; Holley, C.; Thomalla, F. Adaptive governance and managing resilience to natural hazards. *Int. J. Disaster Risk Sci.* **2011**, *2*, 1–14. [CrossRef]
64. Mall, R.K.; Srivastava, R.K.; Banerjee, T.; Mishra, O.P.; Bhatt, D.; Sonkar, G. Disaster Risk Reduction Including Climate Change Adaptation Over South Asia: Challenges and Ways Forward. *Int. J. Disaster Risk Sci.* **2019**, *10*, 14–27. [CrossRef]
65. Ueda, R. "waju" and Its Evolution with Urban Technology—Japanese sustainable community for disaster resilience. *IOP Conf. Ser. Earth Environ. Sci.* **2019**, *294*, 012012. [CrossRef]
66. Mabon, L. Enhancing post-disaster resilience by 'building back greener': Evaluating the contribution of nature-based solutions to recovery planning in Futaba County, Fukushima Prefecture, Japan. *Landsc. Urban Plan.* **2019**, *187*, 105–118. [CrossRef]
67. Ahern, J. Urban landscape sustainability and resilience: The promise and challenges of integrating ecology with urban planning and design. *Landsc. Ecol.* **2013**, *28*, 1203–1212. [CrossRef]
68. Kennedy, J.; Ashmore, J.; Babister, E.; Kelman, I. The meaning of "build back better": Evidence From post-tsunami Aceh and Sri Lanka. *J. Contingencies Crisis Manag.* **2008**, *16*, 24–36. [CrossRef]
69. Hartig, J.H.; Detroit, G.; Heritage, A.; Initiative, R.; Kerr, J.K. Promoting Soft Engineering along Detroit River Shorelines. *Land Water* **2001**, *45*, 24–27.
70. Nateghi, R.; Bricker, J.D.; Guikema, S.D.; Bessho, A. Statistical analysis of the effectiveness of seawalls and coastal forests in mitigating tsunami impacts in iwate and miyagi prefectures. *PLoS ONE* **2016**, *11*, e0158375. [CrossRef]
71. Kay, R.C. Integrated Coastal Planning and Management in Asian Tsunami Affected Countries. In Proceedings of the Workshop on Coastal Area Planning and Management in Asian Tsunami-Affected Countries, Bangkok, Thailand, 27–29 September 2006; FAO Regional Office for Asia and the Pacific: Bangkok, Thailand, 2007; pp. 229–252.
72. Takagi, H.; Mikami, T.; Fujii, D.; Esteban, M. Mangrove Forest against Dyke-break induced Tsunami in Rapidly Subsiding Coasts. *Nat. Hazards Earth Syst. Sci. Discuss.* **2016**, *16*, 1629–1638. [CrossRef]

73. Srinivas, H.; Nakagawa, Y. Environmental implications for disaster preparedness: Lessons Learnt from the Indian Ocean Tsunami. *J. Environ. Manag.* **2008**, *89*, 4–13. [CrossRef]
74. Shibayama, T.; Esteban, M.; Nistor, I.; Takagi, H.; Nguyen, T.; Matsumaru, R.; Mikami, T.; Ohira, K.; Ohtani, A. Implicaciones del tsunami de Tohoku del año 2011 para la gestión de desastres naturales en Japón. *Obras Proy.* **2012**, *11*, 4–17. [CrossRef]
75. Sathiparan, N. An assessment of building vulnerability to a tsunami in the Galle coastal area, Sri Lanka. *J. Build. Eng.* **2020**, *27*, 100952. [CrossRef]
76. United Nations Office for Disaster Risk Reduction. *Global Assessment Report on Disaster Risk*; UNDRR: Geneva, Switzerland, 2022.
77. Sillmann, J.; Christensen, I.; Hochrainer-Stigler, S.; Huang-Lachmann, J.-T.; Juhola, S.; Kornhuber, K.; Mahecha, M.; Reichstein, M.; Ruane, A.C.; Schweizer, P.-J.; et al. *A Briefing Note on Systemic Risk*; International Science Council: Paris, France, 2022.

Article

Three Strategies of Urban Renewal for One National Outline Plan TAMA38: The Impact of Multiparametric Decision-Making on Neighborhood Regeneration

Dalit Shach-Pinsly 

The Faculty of Architecture and Town Planning, Israel Institute of Technology, Haifa 3200003, Israel; dalitsp@technion.ac.il; Tel.: +972-54-4377408

Abstract: The urban renewal of deteriorated areas is a challenge for many city decision-makers. In this study, we aimed to understand the role and impact of the Israeli national outline plan, TAMA38, on urban renewal areas by examining three urban renewal strategies. This plan was developed to strengthen individual buildings against earthquakes, but it also serves as a catalyst for the renewal of deteriorated individual residential buildings in old neighborhoods, particularly in high-demand districts. TAMA38 focuses on the renovation of individual buildings, primarily residential, but neglects the comprehensive vision of the public and private needs of the neighborhood/site complex, of which the individual building is only one component. To understand which planning strategy will achieve better spatial results under TAMA38, a broader examination is required. The objective of this study was to assess the performance of three urban sites developed under the TAMA38 program in the city of Haifa using three main strategies: (1) one comprehensive plan led by one developer with a change in building locations (2) one comprehensive plan but led by diverse developers, while building locations remain unchanged and (3) individual building renewals with no comprehensive plan. The methodology for this analysis was based on the evaluation of various quantitative and quality parameters that influence the performance of the built environment. The results of the research emphasize the need to choose an urban renewal strategy tailored to a specific location, as well as the need for the authority to take responsibility for planning open public spaces throughout the process.

Keywords: urban regeneration; urban renewal; urban analysis; urban evaluation; parameters; neighborhood evaluation

Citation: Shach-Pinsly, D. Three Strategies of Urban Renewal for One National Outline Plan TAMA38: The Impact of Multiparametric Decision-Making on Neighborhood Regeneration. *Architecture* **2022**, 2, 616–636. https://doi.org/10.3390/architecture2040033

Academic Editors: Iftekhar Ahmed, Masa Noguchi, David O'Brien, Chris Tucker, Mittul Vahanvati and Avi Friedman

Received: 21 June 2022
Accepted: 16 September 2022
Published: 20 September 2022

Publisher's Note: MDPI stays neutral with regard to jurisdictional claims in published maps and institutional affiliations.

1. Introduction

Urban renewal has developed over the years around the world based on multiple strategies. In recent years, with technological innovation and understanding of the importance of urban spaces' performance and quality, the matter of which urban renewal strategy is used at each site has become very important. Urban renewal can occur as a top-down process [1], in which the municipal or national authority initiates the urban renewal process of a building or a neighborhood, or as a bottom-up process, in which the residents themselves initiate the renewal process [2]. However, the question that needs to be addressed is what type of built environment is created in the area of urban renewal? Has the built environment's urban quality and performance improved? Which urban renewal strategy can lead to the creation of a higher-quality environment? What urban parameters are important in implementing an urban renewal strategy—for example, who is responsible for the development of open public spaces in site renewal where each building is renewed privately, or what are the execution phases that need to be implemented? The aim of this study is to understand which urban renewal planning strategy for a site produces a better quality and performance of the built environment for the residents who will live in the renewed site and for the neighborhood. The study focuses mainly on the physical–spatial aspect of the open space obtained after urban renewal.

In the state of Israel, urban renewal planning is currently initiated from both directions, top-down planning directed by the authority, or the state and bottom-up planning initiated by residents and the private sector [3]. However, top-down planning entails diverse problems, including difficulty in identifying complexes that are suitable for urban renewal [4] and long renewal processes [5]. On the other hand, the bottom-up renewal process mainly affects individual buildings and not the entire neighborhood. For both urban renewal perspectives, the focus is on the economic and legal aspects and less on the physical and spatial aspects.

The main strategies for urban regeneration focus on demolishing old buildings, constructing new ones, mainly improving the quality of buildings [6], infilling construction in open areas [7], renewing deteriorating urban areas by assimilating new communities [8], etc. However, there is a lack of understanding of the renewal area's quality and performance [9] and which is the better urban renewal strategy in a certain location. This is a significant consideration for the decision-makers, who need to understand how to integrate the free market approach in urban renewal, specifically to address the public sector's need for higher-quality renewed urban environments [10].

Examining strategies for physical urban renewal around the world entails examining diverse approaches. For example, in Hong Kong, the perspective is holistic, as key design factors serve as a basis for sustainable urban regeneration [11]. In Singapore, the focus is mainly related to upgrading existing public housing buildings for more sustainable public housing [12]. In several countries, such as Germany, the focus is on preventing gentrification by developing affordable housing in brownfields and using models of public intervention, with less focus on the physical environment [13]. In the Netherlands and other Western European countries, the strategy focuses on integrated, smaller-scale neighborhoods with the involvement of market partners, residents, and the public, based on local urban problems and appropriate policy responses as well as the social mix in a neighborhood [14]. Austria applies the "soft urban renewal" model, which aims to develop affordable housing in mixed-use sites based on improving and renovating the existing urban environment [15]. In the U.S., urban renewal is aimed at transforming large-scale public-housing sites into small-scale mixed-income projects, mainly based on the private sector [16].

The aim of this research is to evaluate the quality and performance of three urban sites, from an urban planning point of view, developed under the TAMA38 plan using three different strategies. Furthermore, we consider the role of the quality of open areas (public, private, green, or paved) in the outcome of a site's urban renewal.

1.1. Evaluation of the Regenerated Site

Because urban renewal projects are large projects, in most cases with large budgets, the economic interests are indeed of high importance, as is the proprietary legal aspect of the renewed space [17,18]. Therefore, most studies on urban renewal deal mainly with the economic and legal aspects of urban renewal and neglect the built environment's spatial and physical aspects [9,19]. In addition, several interests need to be considered in the promotion of an urban renewal project at the city, neighborhood, and resident levels, which entails some conflicts of interest [9]. Physical conflicts also arise, such as construction (private and public) vs open space and public vs private [19]. Currently, there is a lack of methods and tools for the evaluation of built environments and the public open space for urban renewal decision-making, mainly regarding which urban renewal strategy is suitable for a certain location for regeneration. The result of the urban renewal affects the quality of life in the open space in the private and public realms. Therefore, this study focuses on the physical–spatial dimension of the built environment.

A gap in the literature that the authors have identified is the lack of the evaluation of quality parameters in large urban renewal projects. Although there is evidence-based knowledge regarding the evaluation of the environment [9], there is little connection between this knowledge and the implementation of urban renewal projects. Furthermore, the methodologies that researchers have used in previous studies to evaluate urban renewal strategies neglect the importance of environmental quality, particularly open spaces.

1.2. Local Strategies for Urban Renewal

This article focuses on the quality results of the public and private open spaces of urban renewal environments. For the last few years, urban renewal in the state of Israel has been promoted based on several main pathways, top-down, initiated by the municipality or the state, and bottom-up, initiated by the residents, including:

1. *Strengthening an existing structure based on the condensation of construction*: Adding built areas to existing buildings, reflected as additional floor levels, rooms, balconies, and elevators;

2. *Demolition and reconstruction, evacuation–construction:* The demolition and reconstruction of one or more buildings with considerable consolidation. The contractor is required to evict all tenants and provide them with temporary housing at his or her expense. Local authorities or tenants, with the urban renewal administration's support, promote these projects in their municipal areas. As a result, the municipal authority helps residents assess the project's economic viability and points them in the right direction—either private contractors or public authorities that will benefit from a reduced tax burden;

3. *National outline plan TAMA38:* In 2005, the government approved national outline plan No.38 to strengthen buildings against earthquakes. The plan's purpose is to encourage building residents to strengthen their residential buildings against earthquakes by creating an economic incentive to utilize the building rights [20]. In this plan, a building that was built before 1980 and does not meet standard building regulations can be strengthened and apartments can be expanded by adding additional room areas, closing open ground floors, and adding floor levels. Each TAMA38 project is promoted separately by each building's residents or by private entrepreneurs chosen by the residents [17,18]. TAMA38 is an Israeli national plan that allows construction permits to be issued without a detailed outline plan [21]. This plan has three main routes: (1) reinforce and strengthen the building to increase its earthquake resistance, adding an additional floor, and improving rooms (TAMA38/1); (2) the demolition and reconstruction of the building, including adding 2.5 floors and enlarging the residents' apartments (TAMA38/2); (3) strengthening the building and adding 2.5 floors (TAMA38/3). The plan focuses mainly on extending construction rights, namely for additional residential units on the existing roof, which is the contractor's incentive to finance the construction reinforcements.

It should be noted that since TAMA38 was approved, it has affected many plans that have been promoted in various areas, mainly by increasing the construction sqm (adding more rooms, floor levels, etc.). TAMA38 has a significant impact on the built environment, mainly on the open space, public and private. In this study, we examine three strategies for urban renewal for urban sites based on three perspectives: (1) one comprehensive plan, one developer with a change in building locations; (2) one comprehensive plan, diverse developers and no change in building locations; and (3) no comprehensive plan, individual building renewals, without changing building locations. The TAMA38 plan inspired all the projects.

There is a growing social need for large-scale physical urban renewal projects with the significant involvement of the public sector [22]. Therefore, in this study, we aim to understand which urban renewal strategy will yield the best contribution to the built environment for the residents and for the public and in relation to the renewed site's quality and performance.

1.3. Background of Urban Renewal Strategies

Strategies for urban renewal cover a variety of topics and have recently focused on future planning and connection to innovative trends in urban planning. The relationship between urban renewal and ICT (information and communication technology) has been examined for the last few years. Benkő [23] examined the role of ICT in urban planning and design as a contributor to urban renewal. Their results show the need to develop new methodologies that need to be integrated with traditional top-down planning and design to solve planning problems that affect the built environment. Das [24] argued that cities in the Global South can be revitalized using ICT, especially during urban renewal processes. He argued that ICT in the city centers of countries in the Global South can assist urban renewal in many essential city activities, for example, by monitoring crime for better and livable cities. An additional research route related to the impact of the smart-city approaches to social, economic, and spatial planning is the use of digital questioning about urban renewal strategies for more sustainable cities [25]. Their findings show that emphasizing digital urban renewal with smart cities can be significant for adopting relevant strategies and policies for future planning.

Digital placemaking strategies have been developed as an emerging concept for the renewal of public open spaces in cities [26]. Shih [27] developed space matrixes for digital placemaking by identifying spatial areas with significant potential for digital placemaking and argued that digital placemaking generates a "hybrid space" between the digital and the physical worlds and expands the way people can experience spatial environments. Ioannou [26] identified focal places in the public open space based on a digital placemaking platform. They showed that information about the public open space that is frequently updated using social media provides an updated and accurate picture of an area and helps change perceptions of public spaces.

1.3.1. A Review of Urban Renewal Strategies

Several researchers have argued that urban renewal strategies need to be analyzed together with sustainability. Zheng [28] conducted an integrated review of urban renewal strategy, planning, and sustainability together for the first time, focusing on the social and planning sub-system of urban renewal in terms of assessing sustainability. Based on an analysis of 81 articles on sustainable urban renewal from the period 1990–2012, they pointed out the complexity of achieving sustainable urban renewal and understanding the sustainable mechanism behind the urban renewal process. For the last decade, several approaches to evaluating urban renewal strategies in the built environment have been developed, which address diverse topics. For example, Zheng [29] proposed a framework for evaluating neighborhood sustainability for better urban renewal decision-making for high-density cities based on a decision-making matrix of urban regeneration strategies (such as social aspects, economy and work, resources and environment, and land-use form) and building condition (such as building age and materials). They show that when building conditions and sustainability have high values, it is necessary to conserve the neighborhood. Later, the decision-making matrix was extended to provide implementation paths for urban renewal at the neighborhood level. The matrix is based on neighborhoods' specific problems and characteristics, such as facilities, building conditions, land-use forms, and social, economic, and environmental aspects. After applying this framework in a Chinese neighborhood, decision-makers were able to adjust practical approaches based on the decision-making matrix for small-scale urban renewal improvements in diverse areas of the city [30].

Some consider the city an inclusive natural ecosystem of urban areas, as urban renewal plays a critical role in the neighborhood's life. Ho [31] developed the Dilapidation Index (DI), a structured building assessment scheme for evaluating the suitability of various urban renewal strategies for diverse buildings. The results suggested that management factors and a building's physical conditions play a critical role in differentiating dilapidated buildings from well-performing buildings. They suggest that the DI can help improve the quality of the built environment in urban areas by identifying problematic buildings that impact their surroundings and influence future renewal. Tarani [32] argued that urban creative activities are the ecosystem of spatial concentration. The author refers to the creative activities as cafés, bars, restaurants, etc., and hybrid characters such as creative art spaces and various artist workshops as the organism that develop evolutionary networks and creative interactions that create the collaboration between entities. A bottom-up development, which is a spontaneous phenomenon, happens in diverse urban areas.

Other researchers have argued that the main role of urban renewal development is the renewal of the public open space. Van Melik and Lawton [33] analyzed the role of urban public spaces as an important tool for urban renewal strategies. The authors argued that city decision-makers recognize the importance of public open spaces to the local context and attract private entrepreneurs to invest in improving urban open spaces as part of their planning agendas in various cities. Recently, several researchers have aimed to understand renewed neighborhoods' impact on the residents relocating from their old neighborhoods. Miltenburg and others [34] studied the differences between similar types of residents based on statistical data on those who relocated voluntarily and involuntarily and found that there is no conclusive evidence showing that housing relocation leads to more socioeconomic and employment opportunities for those forced to relocate. However, the findings show that forced relocates are living in lower qualitative neighborhoods after relocation.

An additional aspect is the link between urban development and developers' engagement, especially the agreements between developers and planning authorities, that influence construction procedures [35]. The phenomenon of agreements is universal and reflects the trend toward privatization that is prevalent today, in which the private sector takes part in the provision of public tasks, as occurs, for example, in the Netherlands [36], Germany, and the Baltic countries [37]. Developers are harnessed to supply public tasks through agreements with the planning authorities during the preparation of plans [35,38]. In many countries in Europe, authorities are looking for tools to guide transformations in urban renewal by promoting more efficient land use for improved public-value capture [38]. In Israel, the use of a levy tool is more common in municipalities [35]. In Israel, although there is no unique legislation for this field, many authorities usually make agreements with entrepreneurs, mainly in large projects (in the scope of hundreds of units), for different purposes, such as preserving buildings, developing physical infrastructure, improving traffic systems, and developing or promoting public spaces, among others [35].

1.3.2. Assessing the Built Environment's Quality after Renovation

Around the world, green, sustainable evaluation tools have been developed, such as BREEAM, LEED, Green Star Community, and SI 5281 in Israel. The BREEAM (Building Research Establishment Environmental Assessment Method), established by the UK Building Research Establishment (BRE), measures best practices in environmental design and management [39,40] based on metrics of sustainability and indicators that consider health and well-being, management processes, ecology, waste, and more. It currently also focuses on neighborhood development [41,42]. In 2003, the Green Star Community was established in Australia as a sustainable rating system for buildings [39]. A Green Star Community rating can be obtained during the planning and design process. In addition to residents' health, productivity, and operational costs, the rating system is also considered.

As a basis for green building practices, LEED (Leadership in Energy and Environmental Design, which the U.S. Green Building Council, USGBC, developed) is accepted nationally [42,43]. LEED consists of rating systems for the design, construction, operation, and maintenance of buildings [44] and currently serves as the main tool for evaluating sustainability in the U.S. Furthermore, the LEED-ND for neighborhoods was developed, which incorporates principles of smart growth, urbanism, and green building into a national rating system for neighborhood design in the U.S. It recognizes urban projects that enhance overall the health, natural environment, and quality of life through pedestrian-friendly neighborhoods, public transportation, and green buildings and infrastructure [45]. For the last few years, a rating-based system has been developed in Israel, "The neighborhood-360°". This measurement system is used to promote sustainable neighborhoods based on high-quality, healthy, and livable development and construction. The tool is based on three main elements, public and natural spaces, construction and infrastructure, and the efficient use of resources, which allow for the assimilation of quality principles of design into the built environment through the integration of multidimensional building and development challenges by using the quality evaluation criteria of the planning and design process [46].

Recently, attempts have been made to evaluate the built environment using innovative techniques, such as multiparametric analysis to evaluate alternatives for urban regeneration [9]. Shach-Pinsly and others [9] developed a 3D-GIS multiparametric evaluation analysis to evaluate the quality and performance of the urban environment as a significant result of the urban renewal decision-making process at the neighborhood level. Furthermore, the concept of "performance-based codes" [47] helps us understand the role of quality and performance during the planning and design process. This approach allows for the assessment of various scenarios during the urban renewal process.

2. Methodology Framework

The aim of the research was to understand how three urban renewal strategies develop different urban environments under the same "national outline plan", TAMA38. The study's methodology is based on understanding the performance and quality each urban renewal strategy yields by assessing several environmental quality parameters and evaluating the renewed built environment, including:

1. Quantitative parameters, among them, public and private open space size, residential density, etc;
2. Quality parameters, among them, public participation, walkability, the examination of the execution phases, etc;
3. Examining the resulting sustainability of the three urban environments before and after the urban renewal process, considering the renewal changes.

Quantitative parameters can be measured objectively by size, area, number of residential units, and other factors. Quality parameters refer to urban parameters that reflect the site's quality, before and after urban renewal. Urban planners and designers can evaluate these parameters by understanding whether a public participation process has been implemented or whether the phases of the design process have been analyzed by the authorities. Alternatively, using suitable methods and tools, urban planners and designers can measure parameters such as walkability, visibility, and others.

The project area selection was based on existing masterplan projects approved by the city municipality, which revealed the need for further evaluation and analysis to understand the evolved quality of the renewed, developed environment. We aimed to assess the performance of three renewed sites in the city of Haifa: Berl-Katzenelson, Haviva-Reich, and De-Israeli. The main urban renewal in each site occurred along a main street.

We developed a research flow, which included the following phases (Figure 1):

a. Analyze three urban renewal strategies based on the same planning policy of "national outline plan" TAMA38;
b. Analyze each renewal strategy and the renewed site according to environmental quality/performance criteria;
c. Conduct a comparative evaluation between the three urban renewal strategies and cases studies;
d. Urban renewal decision-making based on the TAMA38 planning strategy—develop a basis for decision-making for urban renewal based on the analysis of the three urban renewal strategies.

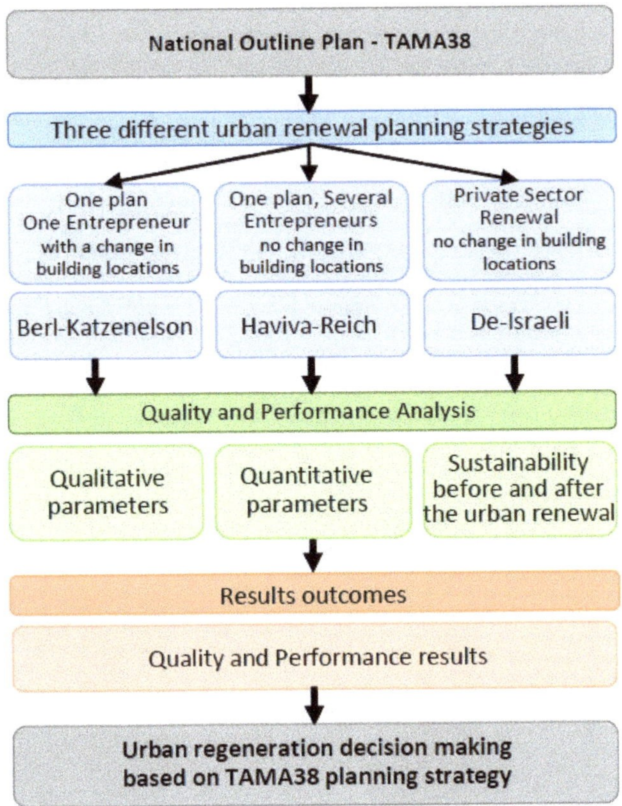

Figure 1. The study's flow.

We designed the study from the planners' point of view and at the neighborhood-built environment scale, not from the individual architectural-building scale, considering the decision-making process. Our main goal was to understand each urban renewal strategy's *advantages* and *disadvantages* and understand which parameters and aspects can be added for a better decision-making process for a renewed site. Therefore, the selection of analyzed parameters and adjusting them to the neighborhood scale is important for the urban renewal strategy evaluation.

Introducing the Case Studies: Berl-Katzenelson, Haviva-Reich, and De-Israeli Sites

All three case studies are located in the city of Haifa, in three different areas of the city. All case studies are adjacent to one of the city's commercial centers. Following is a description of the case studies before and after the renewal process, including various aspects affecting the regenerated sites' performance:

Berl-Katzenelson Site:

Before the renewal process: The Berl-Katzenelson site is located in the southwestern part of the Neve Sha'anan neighborhood, adjacent to a wide public open area, and very close to the city's northern commercial centers. Richard Kaufman (architect) founded the neighborhood around the 1920s. The site is located along a one-way street and is relatively flat, with one entrance and one exit, and is very walkable. The neighborhood is relatively close to the Technion-ITT. The Berl-Katzenelson site (the area for renewal) includes seven buildings that are three or four stories high, with 126 small 2–3-bedroom housing units developed during the 1950s. The housing density is relatively low, at

4 apartments per dunam (1000 sqm). The parking spots were all public, spread along the street, with no private parking inside the building plots. Most of the street population were elderly, students, and ultra-Orthodox, who opposed the process of approving the Urban Renewal Berl-Katzenelson program. This ultra-Orthodox population benefited from a row of buildings located two floors above the entrance bridge and two floors below the entrance that provided access to apartments without the use of elevators and that helped preserve the ultra-Orthodox way of life. The neighborhood's statistical cluster level is 6 (out of 10). All infrastructures are outdated and crumbled.

Urban renewal strategy: The renewal plan, Plan HP/2281, was approved in 2011. Building layouts and locations were different from existing plans. The renewed site includes seven renewed buildings, four of which are fifteen stories high and three nine stories high, for a total of 370 resident units. The construction began in 2018. However, when the construction began, the number of resident units increased to 484, more than tripling the old unit's density. The construction includes parking lots and ancillary services for the tenants' well-being, such as a sports club, a tenants' club, and transformation rooms, all within the private plots. Additional public parking was developed along the street, and additional routes and parking were added at the bottom of the site. The wild public open area around the building was reduced, and a designed open area was developed between the buildings and up the street. The part of Berl-Katzenelson Street included in the developed site was transformed from a one-way street to a two-way street (Figures 2 and 3). With this strategy, the residents took part in a public participation process.

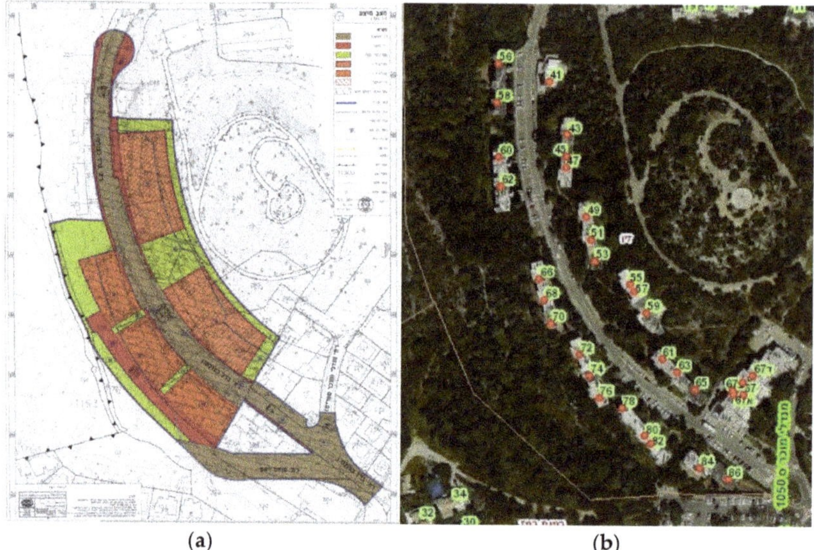

(a) (b)

Figure 2. (**a**) Berl-Katzenelson Site, the approved plan, 2006. Reduction of the public spaces between the buildings (orange: residential housing; green: open areas; red: new road). The design: Uzi Gordon Architects. Source: the Israeli "Available Planning" website (The Planning Administration, Ministry of Interior) archive; (**b**) Berl-Katzenelson Site, existing situation before the urban renewal. Source: the city of Haifa website, 2015.

(a)

(b)

Figure 3. Berl-Katzenelson Site. A view from the inner side of the site (**a**); A view from the entrance to the site (**b**). Photographed by the author, 2022.

Haviva-Reich Site:

Before the renewal process: The Haviva-Reich site is located in the heart of the Ramot-Remez neighborhood, which was developed during the 1950s and 1960s. The site is within walking distance of one of the northern city's significant commercial centers and lies between two major universities, the Technion-IIT (Technion-Israel Institute of Technology) and Haifa University. The renewed area comprises 15 buildings located along one main rounded street, with one entrance and two exits. A natural forest surrounds the area, at the center of which lies a green nature reserve area. The area was developed along a slope, partly steep and partly moderate. The buildings are three to four stories in height and include around 330 small housing units of identical two-bedroom apartments. Between the buildings are wide-open green areas of around 20 m and longer. The housing density is five units per acre. The parking spots are public, spread along the street, with no private parking inside the building plots. The area's walkability is comfortable, with sidewalks, leading to main streets, a neighborhood center, schools, and universities. The main rounded street, Haviva-Reich Street, is served continually by three bus lines, several times an hour throughout the day. The residents are mostly elderly and young students. The neighborhood's statistical cluster level is 6–7 (out of 10). All infrastructure is outdated and crumbling [19].

Urban renewal strategy: The renewal plan includes fifteen buildings eight or nine stories tall located along Haviva-Reich Street, in a similar layout and location of the demolished buildings (Figures 4 and 5). The open area between buildings was reduced and left with one large open area at the center of the neighborhood. The developed plan comprises residence units and several commercial areas. The first building was developed under a system of reinforcement against earthquakes by extending the original apartments and adding three floors to the roof, doubling the number of housing units. The masterplan was updated in 2007 when the development mechanism changed to demolishing the old buildings and building new ones. This plan enabled the addition of five to six more floors and up to 27 m of height as well as the tripling of the units' density, resulting in around 930 residential units. The number of parking units was reduced to one per new apartment, and the public parking along the street was reduced. Although the site was developed as one plan, each building was developed by a separate entrepreneur, and their profits come from the sale of the new apartments [19]. In this strategy, the residents were not involved in any public participation process; furthermore, the site's residents objected to the design plan [19].

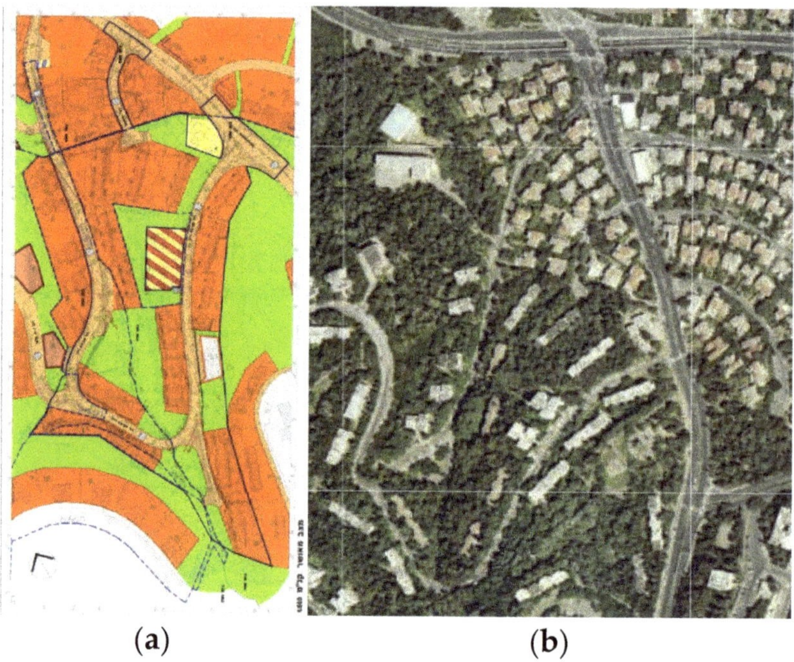

(**a**)   (**b**)

Figure 4. Haviva-Reich site, the approved plan, 2006. Reduction of the public spaces between the buildings (orange: residential housing; green: open areas; brown: public buildings). The design: Giora Gur Architects. Source: the Israeli "Available Planning" website (The Planning Administration, Ministry of Interior) archive (**a**); Haviva-Reich site after the urban renewal Source: govmap, 2007, https://www.govmap.gov.il/ (accessed 15 June 2022) (**b**).

De-Israeli Site

Before the renewal process: The De-Israeli site is located in the heart of the Carmel neighborhood, which was established around the 1920s, mainly for wealthy family homes. The neighborhood is located adjacent to one of the city's commercial centers, which constitutes the southern end of the municipal business center of Carmel ridge in Haifa. The De-Israeli site is located along a one-way street that is relatively flat and has one entrance and one exit, with old trees along the sidewalks. The buildings were developed individually (under the outline plans of the City of Haifa) over the years, mainly before 1990. The buildings are around four or five stories tall, and they include diverse three- and four-bedroom apartments. The De-Israeli site includes around 550–600 housing units along the entire street, with a density of approximately 6 units per dunam (1000 sq. m.), which is a medium-to-low density. The site's public parking spots are spread along the street, and some of the building lots include several private parking spaces. The housing residents are diverse, mainly families and the elderly. The neighborhood's statistical cluster level is 7–9 (out of 10). The infrastructure is partly outdated and crumbling. Two elementary schools are located at one end of the street (a religious state school and a private school), as are several types of freelancer offices.

Figure 5. Haviva-Reich site, A view from the inner side of the site (**a**); A view from the entrance to the site (**b**). Photographed by the author, 2022.

Urban renewal strategy: The renewal planning is based on the TAMA38 plan (approved in 2005), a national outline plan that supports old buildings' residents in strengthening their buildings against earthquakes by providing economic incentives for extensive construction development, such as adding residential units, balconies, rooms, and parking areas (Figures 6 and 7). Each building's renewal development is managed and constructed individually, detached from the adjacent buildings, with no comprehensive masterplan for the site and no municipal intervention in the public open area. The TAMA38 plan's potential for the De-Israeli site is to double the number of residential units. However, because there is no comprehensive masterplan and each building's residents decide individually regarding the building's renewal, the number of housing units at this time is dynamic and growing. Moreover, there is minimal municipal intervention in the development of the public space, and no public or commercial buildings have been added to this site. In this strategy, the residents are provided a public participation process in the form of a standard objection process conducted separately for each building.

(a) (b)

Figure 6. De-Israeli site before the urban renewal (**a**) and after the urban renewal (**b**). Source: (**a**) govmap, 2015; govmap, 2021, https://www.govmap.gov.il/.

Figure 7. De-Israeli site, during the transformation of TAMA38 construction. A view from the inner side of the site. Photographed by the author, 2022.

All three sites have many similar geographical features: secondary one-way streets, a circular street with one entrance and one exit (or two), located at a walkable distance from a significant urban center. However, the three urban renewal strategies led to the development of three distinct urban environments.

3. Analysis Results

To evaluate the quality of the three urban renewal strategies, we defined several parameters (31), based on available data, and models and tools for evaluating the criteria for the analysis. We reviewed diverse parameters for the analysis and selected those that met the urban renewal analysis criteria before dividing them into two groups: The first group includes *quantitative parameters* (Table 1), including the starting year of the masterplan, urban renewal strategy and track (municipality), new masterplan and year of approval, process starting and ending years, number of existing residential units, number of approved residential units (first phase, at the first approval stage of the plan), residential units approved (for development permit; received building permission), number of entrepreneurs for the entire site (for the three case studies), responsibility for public open-space development (who is responsible for the open areas' development and maintenance), parking regulation, and public transportation. The second is *urban environmental quality parameters* (Table 2), including resulting residential density in the renewed site, changes in public open areas and green areas, responsibility for the development of public open spaces, public open spaces (in relation to the area), private open spaces (in relation to the area), vegetation (reduction or incision), changes in the road network, changes in public transportation (number of buses and frequency), private transportation, traffic load, parking (reduction or incision), bicycle path, walkability (measure by map whether the walkability is improved), visibility (the views from the buildings and from the street level [48]), new public building development, access to public and commercial services, mixed-use, public participation (discovering the process), change in population identification (gentrification), and execution phase examination. The tools for analyzing the quality and performance aspects were validated in previous research [9], including visibility [48] and walkability [49]. Table 1 presents the findings for the basic parameters and Table 2 presents the findings for the urban environmental quality parameters.

Table 1 shows the differences between the three urban renewal strategies in relation to the quantitative parameters. The evaluation indicates the basic differences between the three strategies as the developed plan, the number of entrepreneurs for the entire site, or the starting year for the masterplan. However, it also shows several similarities between the strategies but with different local results due to the differences between locations and urban renewal strategies. One example is the number of approved residential units (with development permits) that triple the residential units (relatively) in each site followed by a significant increase in the density of the site. Another example may be the similarity of the parking standard that results in parking density in the public street areas for all strategies. A comparative evaluation of the quantitative parameters indicates that differences in the results of these parameters can reflect similarly on the outcome of an urban renewal project, as in similar quantitative parameters.

Table 2 shows the differences between the three urban renewal strategies in relation to the quality parameters at several levels. The comparable evaluation indicates a decrease in quality between the three strategies in relation to open public and private spaces and vegetation. The evaluation indicates no change in several qualities, such as the road network, public transport, bicycle paths, walkability, mixed uses, and access to public and commercial services. However, the evaluation indicates increased levels in several parameters that influence the quality of the site, including traffic load, public parking, and visibility at lower levels. Several parameters point out aspects that may have a stronger influence on levels of quality: the municipality's responsibility for the development of open public spaces, public participation, the execution of examinations for phases, and the additional development of new public buildings. These parameters have a wide impact on the overall quality of the urban renewal complex depending on their implementation in various planning stages.

Table 1. Comparable evaluation of quantitative parameters between the three case studies.

	De-Israeli Site	Berl-Katzenelson Site	Haviva-Reich Site
Starting year of the masterplan	2005 (TAMA38)	2006	2005
Plan and track goals. Urban-renewal strategy	Urban renewal based on TAMA38 national plan	Urban renewal based on comprehensive masterplan; the authority's route	Urban renewal based on comprehensive masterplan; the authority's route
New masterplan and year of approval	No inclusive masterplan; TAMA 38; approved in 2005	HP/2281 Urban Renewal- Berl-Katzenelson approved in 2011	HP/2187 Urban Renewal- approved in 2006
Starting/ending year of the process	No masterplan; continued development at request of private sector	Beginning of site work in 2018; plan implementation expected to reach 100% in 2023	Beginning of site work in 2013; plan implementation expected to reach 100% in 2021
Number of existing residential units	550–600 residential units along entire street; 6 units per 1000 sq. m. (one dunam)	126 units	330 units
number of approved residential units (first phase)	According to TAMA38, double the number of residential units	372 units	572 units
number of approved residential units (with development permits)	According to TAMA38, additional residential units based on request for construction relief and approved by authority	484 units	930 units
Number of entrepreneurs for entire site	Related to number of TAMA38 development permits	One entrepreneur	Close to 10 entrepreneurs
Responsibility for public open spaces development	Haifa Municipality's responsibility	Entrepreneurs' responsibility, except the main wide open public green space	Haifa Municipality's responsibility
Parking standard	1.5 parking spaces per apartment and a third parking space per apartment for guests	1.5 parking spaces per apartment and a third parking space per apartment for guests	One parking space for one new residential unit; no new parking spaces for existing/old units or for guests
Public transport	One bus line, one bus line per hour, throughout the day	One bus line, one bus line per hour, throughout the day	Three bus lines, one per hour for each throughout the day

Table 2. Presents the comparable evaluation quality parameters for the urban environmental quality parameters.

	De-Israeli Site	Berl-Katzenelson Site	Haviva-Reich Site
Residential density	Increased	Increased	Increased
Changes in public open areas and green areas	No change in open public green areas	Many public green open areas have been transformed into new building lots, a new street, and new parking areas.	Many public green open areas have been transformed into new street areas and new parking areas.
Responsibility for development of public open spaces	Haifa Municipality's responsibility	Entrepreneurs' responsibility except for the main wide-open public green space	Haifa Municipality's responsibility
Public open space	No change	Reduced	Reduced
Private open space	Reduced per building	Reduced	Reduced
Vegetation	Reduced	Reduced	Reduced
Changes in the road network	Local street changes, with no change in the road network	Local street changes, with no change in the road network	Local street changes, with no change in the road network
Change in public transport	No change in route or availability	No change in route or availability	No change in route or availability
Private transportation	Increased	Increased	Increased
Traffic load	Transport load increased	Transport load increased	Transport load increased
Parking	Inside private lots, reduced parking area, and lack of public parking	Inside private lots, reduced parking area, and lack of public parking	Inside private lots, reduced parking area, and lack of public parking
Bicycle path	No bicycle path added	No bicycle path added	No bicycle path added
Walkability	No change	No change	No change
Visibility	Reduced for lower levels and increased for upper levels	Reduced for lower levels and increased for upper levels	Reduces for lower levels and increased for upper levels
New public-building development	Lack of reference to public buildings; no additional public-building design	Lack of reference to public buildings; no additional public-building design	Lack of reference to public buildings; no additional public-building design
Mixed use	No change	No change	No change
Access to public and commercial services	No change	No change	No change
Public participation	No public participation	The residents generated the initiative for urban-renewal process, so public participation took place throughout the process.	No public participation

Table 2. *Cont.*

	De-Israeli Site	Berl-Katzenelson Site	Haviva-Reich Site
Change in population setting (gentrification)	Additional residents for each separate building	Most residents, including those who for many years could not live in the complex because of its condition and were forced to live in other apartments, are expected to return to their homes.	A gentrification process occurred with the arrival of young families and children with high socioeconomic profiles.
Execution phases examination	Individual development with execution phases according to residents' abilities	Local authority and entrepreneur control project's sales execution progress	Lack of execution phases and authority control

4. Discussion

The three urban renewal projects presented in this study demonstrate the difficulties and successes of urban renewal processes as well as their complexity. Currently, the urban renewal process in the State of Israel is mainly focused on the economic aspect and the supply of new apartments to the housing market [21]. However, as studies have shown, this goal is characterized by many problems, including social problems, transportation problems, land-use problems, and severe damage to nature, which are not addressed in the current urban renewal planning policy. Therefore, there is a need for a change in the main goal of urban renewal, which should enhance the renewed area's public open space according to the existing population's needs and desires. The three case studies' analysis results revealed several recommendations and improvement suggestions for the urban renewal process of the site and the neighborhood, and they highlighted diverse problems that emerged as insights from the analysis. Following is a summary of the insights and a presentation of the improvement suggestions.

The need for planning with a comprehensive vision—in preparing the masterplan, an inclusive vision is necessary that includes the area surrounding the urban renewal site and its constraints. It should ensure the maximum utilization of the land's resources while maintaining the existing urban fabric, with optimal planning tailored to the neighborhood. An inclusive vision should be implemented in demolition and reconstruction sites as well as in areas where TAMA38 is implemented. Another goal is to renew the infrastructure and landscape, among other elements, and avoid disruption to the residents' site.

Combining old and new development—additional construction that can be approved for existing buildings may damage a building's exterior and cause old and new buildings to clash in appearance. Extensive construction additions to existing old buildings can be a planning and design challenge [18]. The goal is to reduce the effect of increasing a building's influence on surrounding buildings while devoting planning to ensure the building's integration into its environment [18].

Building typologies—recommendations to the local authority to produce a catalog of building typologies suitable to build in the renewed site during the urban renewal process, taking into consideration the neighborhood's topography, location, and socioeconomic level. Developers offer building typologies and layout locations based on their experience, sometimes regardless of local situations, and lack tools for preserving old buildings. Additionally, architectural solutions for TAMA38 cannot be produced for all construction-related and residential needs because of constraints.

Preserving existing natural values—cutting down large numbers of trees on behalf of urban renewal is an ongoing problem. The original natural area values cannot be preserved in any of the renewed routes examined. Emphasis should be placed on the existing natural areas during preparation for the urban renewal process. There is a need to integrate existing and planned trees, consider existing water sources, find original solutions to incorporate them into the landscape, and treat these solutions as valuable resources.

Changing the neighborhood characters—urban renewal, including TAMA38, allows for increasing buildings' volume by adding housing units to and expanding existing buildings. Additionally, renewal programs allow for the demolition of old buildings and the construction of new buildings by expanding the structure contour in relation to the lot size. The urban fabric affects changes in the surrounding and distant environment for each renewed building or for a whole site, with no preservation of the previous site's qualities, such as its density and traffic. The municipality promotes strengthening buildings and building additions as incentives for developers, even at the cost of changing the nature of the environment and the neighborhood.

Include predefined execution stages for the renewal plan—in contrast to the advantages of one entrepreneur (the case of the Berl-Katzenelson site) who manages the planning and execution stages, execution vis-à-vis municipality features advantages in the number of entrepreneurs (Haviva-Reich and TAMA38 projects), mainly related to the creation of a competition for architectural quality and construction and apartment prices. However, the most important step is defining execution steps in the plan to avoid disruptions to the local residents' daily routines.

Construction density—although the main objectives guiding local authorities' urban renewal processes are the economic and legal aspects, the solution of increasing the density multiplier is not suitable for all urban renewal projects. Therefore, there is a need to include the social and physical aspects. Moreover, additional residential units in the existing urban fabric may burden the road system and cause traffic congestion. Our recommendation is to define a mechanism for determining the multipliers for the number of residents in accordance with the type of urban renewal route, the neighborhood's size, geographical and topographic location, land value, market price levels, and other factors. A process is necessary to establish criteria that include the designer's and contractor's competitive parameters and allow for the selection of a designer and contractor based on the environment's target quality after renewal.

Gentrification and change in population composition—the solution for gentrification must include the authorities. It is necessary to provide solutions for diverse populations that enable all residents to return at the end of construction.

Public buildings and mixed-use—the process of urban renewal requires the construction of additional public buildings. Therefore, it is necessary to include areas for public construction in the renewed masterplans.

Administrative division of urban renewal—the administrative division of urban renewal is a unit of the local authority for promoting urban renewal processes, simplifying these processes, making information accessible to residents, assisting professionals in the urban renewal field, and bridging the gap between the local authority and the residents to accelerate urban renewal processes. A comprehensive vision is necessary to influence the urban renewal plan and process, regardless of which route will be taken.

Public participation—public participation is highly important for urban renewal processes. The process should include relevant professionals (planners, social workers, legal advisers, etc.), representatives of the local authority, entrepreneurs, local residents, and any other party required. In many cases, tenants attempt to organize and work together with the authority, which works poorly because of their lack of knowledge, the apartment owners' professionalism, and a lack of trust in the local authority. Therefore, it is recommended that municipalities incorporate social counselors to advance information processes even before urban renewal procedures begin.

Determining the road network—the analysis shows that the three case studies neglected planning regarding road networks and required treatment at the intersections leading out of the neighborhoods. In all cases, the main roads remain single-road streets, each with a single entrance and a school. The renewal plans neglect the increases in the sites' populations and in the numbers of vehicles in the neighborhoods. Renewal plans need to include the redefinition of road networks with clear definitions of secondary and main streets to enhance road networks' performance following the changes in the population. Furthermore, there is a need to ask whether these locations will be able to adequately address the increased demand created by demographic changes. It should be noted that none of the three routes addressed the transportation problems, and it is impossible to determine whether one route is more effective than the others. Following this structure, the walkability did not change or increase in any of the three case studies.

Public transportation—urban renewal and public transportation must be interconnected. Neglecting this integration may cause traffic problems in the renewed area, as the three case studies show.

Parking standard and development plan—parking needs to be reflected in renewal plans to ensure that streets are clear of cars for the residents' benefit. The design needs to address walkability, connectivity, and bike lanes inside and outside the neighborhood, reduce parking areas in public open spaces, and develop public transportation. Demolition and reconstruction enable the development of parking inside private lots, including parking facilities that double the parking area, and allow more cars to park inside private lots.

To conclude, one of the most important objectives of urban renewal is to develop public infrastructure throughout the renewal design and development process. Doing so will strengthen the urban renewal site, ensure the high quality of the new buildings, and enhance the renewed area's quality and performance.

5. Conclusions

We compared three urban renewal strategies in which buildings were demolished and rebuilt to improve the old buildings' resistance to earthquakes and renew the neighborhood: The urban renewal strategy for the Berl-Katzenelson site provides a comprehensive solution that combined the private interest with the public interest. The urban renewal strategy for the Haviva-Reich site provides a comprehensive solution that was carried out in a fragmented manner. The urban renewal strategy for the De-Israeli site is based on the TAMA38 plan and involves demolishing and rebuilding each old building separately, with no comprehensive plan. All three case studies show that the demolition and reconstruction of the evacuation–construction urban renewal route the most effectively creates a comprehensive urban renewal process.

TAMA38 is a national program that leads to urban renewal in many cities. Therefore, this research is important beyond the city of Haifa. A thorough understanding of strategies for urban renewal that are based on the TAMA38 plan can also influence the development of similar urban renewal strategies, tailored to a location, in other cities in Israel to additional cities in different locations in the world. Haifa is a hilly city with a high demand for cars; therefore, additional parking in open public spaces is needed. However, in this research, we identified several other parameters that are important to implement in strategies for urban renewal and in different areas of the country, such as a municipality's responsibility for the development of public open spaces and the examination of the phases of execution, leading to the promotion of urban renewal based on the TAMA38 plan.

There is no doubt that the economic aspect is highly important, but at the same time, the choice of urban renewal strategy cannot ignore the planning of public open spaces and the built environment as a comprehensive environment, as well as many other considerations mentioned above, which need to be taken into consideration before the urban renewal process begins. It is necessary for the local authority to determine the urban renewal policy in relation to the developers and residents as early as possible. Furthermore, the authority needs to take responsibility for the planning of the public open spaces in all phases, from planning to settling, the completion of infrastructure planning, and work on the open space, including dealing with the developers and contractors.

Nevertheless, it is necessary to achieve a balance among desires to renew private buildings, preserve private properties, and upgrade public open spaces. TAMA38 was developed in 2005 and will be completed around 2023, mainly due to the focus on the buildings and the limited attention directed toward the public open spaces. Urban renewal processes entail many opportunities to improve the urban environment and enhance its quality.

The current research bears several limitations, for example, there is a lack of knowledge regarding the residents' satisfaction (existing and newcomers) with the three sites of their renewed neighborhoods. The residents' satisfaction can affect the behavior of the residents in the neighborhood. The topic is very important and affects the way the residents use the site area, so future comprehensive research would need to include the residents' satisfaction as an important aspect of perpetual research. An additional limitation of this study is the data and information it is based on—public information available for all three strategies and case studies. Future research needs to take into consideration the outcomes of this research while selecting the urban renewal strategy of an urban site. It is necessary to develop a comprehensive urban renewal model based on the TAMA38 plan that incorporates open public spaces and uses for the benefit of the urban renewal site, its neighborhood, and the surrounding areas.

To conclude, there are many varied constraints, conditions, circumstances, and considerations to remember in the selection of the best concepts and strategies for urban renewal processes. Therefore, it is very important to decide which urban strategy is the best tailored to a specific site in a specific location in a city as well as to the residents to improve the development of urban renewal strategies for future urban renewal.

Funding: This research received no external funding.

Institutional Review Board Statement: Not applicable.

Informed Consent Statement: Not applicable.

Data Availability Statement: Copyright data are available under consent from Haifa municipality, and interview results are available from the authors.

Conflicts of Interest: The authors declare no conflict of interest.

References

1. Palumbo, M.L.; Fimmanò, D.; Mangiola, G.; Rispoli, V.; Annunziato, M. Strategies for an urban renewal in Rome: Massimina Co_Goal. *Energy Procedia* **2017**, *122*, 559–564. [CrossRef]
2. Della Spina, L.; Giorno, C.; Galati Casmiro, R. Bottom-up processes for culture-led urban regeneration scenarios. In *International Conference on Computational Science and Its Applications*; Springer: Cham, Switzerland, 2019; pp. 93–107.
3. Ministry of Construction and Housing Website. Available online: https://www.gov.il/he/departments/ministry_of_construction_and_housing (accessed on 1 May 2012).
4. Porat, I.; Shach-Pinsly, D. Building morphometric analysis as a tool for urban renewal: Identifying post-WWII mass public housing development potential. *Environ. Plan. B-Urban Anal. City Sci.* **2019**, *48*, 248–264.
5. Wang, H.; Shen, Q.; Tang, B.S.; Lu, C.; Peng, Y.; Tang, L. A framework of decision-making factors and supporting information for facilitating sustainable site planning in urban renewal projects. *Cities* **2014**, *40*, 44–55. [CrossRef]
6. Carmon, N. Housing policy in Israel: Review, evaluation, and lessons. *Isr. Aff.* **2001**, *7*, 181–208. [CrossRef]
7. Newton, P.; Glackin, S. Understanding infill: Towards new policy and practice for urban regeneration in the established suburbs of Australia's cities. *Urban Policy Res.* **2014**, *32*, 121–143. [CrossRef]
8. Kleinhans, R. Social implications of housing diversification in urban renewal: A review of recent literature. *J. Hous. Built Environ.* **2004**, *19*, 367–390. [CrossRef]
9. Shach-Pinsly, D.; Bindreiter, S.; Porat, I.; Sussman, S.; Forster, J.; Rinnerthaler, M. Multi parametric scenario analysis for urban renewal decision making processes. *Urban Plan.* **2021**, *6*, 172–188. [CrossRef]
10. Hastings, E.M.; Adams, D. Facilitating urban renewal: Changing institutional arrangements and land assembly in Hong Kong. *Prop. Manag.* **2005**, *23*, 110–121. [CrossRef]
11. Deng, Y.; Chan, E.H.; Poon, S.W. Challenge-driven design for public housing: The case of Hong Kong. *Front. Archit. Res.* **2016**, *5*, 213–224. [CrossRef]
12. Teo, E.; Lin, G. Determination of strategic adaptation actions for public housing in Singapore. *Build. Environ.* **2011**, *46*, 1480–1488. [CrossRef]

13. Priemus, H.; Metselaar, G. Urban renewal policy in a European perspective. *J. Hous. Built Environ.* **1993**, *8*, 447–470. [CrossRef]
14. Jing, L.; Sun, L.; Zhu, F. The Practice and Enlightenment of Architectural Renovation and Urban Renewal in the Netherlands. *IOP Conf. Ser. Earth Environ. Sci.* **2020**, *526*, 012200. [CrossRef]
15. Huber, F.J. Chapter 10—Sensitive urban renewal or gentrification? The case of the Karmeliterviertel in Vienna. In *Everyday Life in the Segmented City*; Camilla, P., Gabriele, M., Lorenzo, T., Eds.; Emerald Group Publishing Limited: Bingley, UK, 2011; pp. 223–239.
16. Goetz, E.G. Desegregation in 3D: Displacement, dispersal and development in American public housing. *Hous. Stud.* **2011**, *25*, 137–158. [CrossRef]
17. Bromberg, L. Clearing the way for urban renewal. *Buildings* **2013**, *302*, 8–16. (In Hebrew)
18. Gidron, M.; Namdar, A. *TAMA 38*; Hoshen Mishpat: Jerusalem, Israel, 2012. (In Hebrew)
19. Shadar, H.; Shach-Pinsly, D. From Public Housing to Private Housing: Neglect of Urban Qualities during the Urban Regeneration Process. *Land* **2022**, *11*, 875. [CrossRef]
20. Knesset Israel. 2019. Available online: http://www.knesset.gov.il/mmm/ (accessed on 25 February 2019).
21. Shalev, N. *TAMA38 Plan for Strengthening Buildings against Earthquakes: A Real Solution or a Virtual Solution?* (In Hebrew). Bimkom: Jerusalem, Israel, 2011.
22. Jeffry, P.; Granger, R. Chapter 5—Physical and environmental aspects. In *Urban Regeneration*; Roberts., P., Sykes, H., Granger, R., Eds.; Sage: London, UK, 2016; pp. 87–99.
23. Benkő, M.; Bene, B.; Pirity, Á.; Szabó, Á.; Egedy, T. Real vs. virtual city: Planning issues in a discontinuous urban area in Budapest's inner city. *Urban Plan.* **2021**, *6*, 150–163. [CrossRef]
24. Das, D.K. Revitalising South African city centres through ICT. *Urban Plan.* **2021**, *6*, 228–241. [CrossRef]
25. Praharaj, S. Area-based urban renewal approach for smart cities development in India: Challenges of inclusion and sustainability. *Urban Plan.* **2021**, *6*, 202–215. [CrossRef]
26. Ioannou, B.; Kalnis, G.; Nicolaou, L. Public space at the "palm of a hand": Perceptions of urban projects through digital media. *Urban Plan.* **2021**, *6*, 242–256. [CrossRef]
27. Shih, C.M.; Treija, S.; Zaleckis, K.; Bratuškins, U.; Chen, C.H.; Chen, Y.H.; Chiang, C.T.W.; Jankauskaitė-Jurevičienė, L.; Kamičaitytė, J.; Korolova, A. Digital placemaking for urban regeneration: Identification of historic heritage values in Taiwan and the Baltic states. *Urban Plan.* **2021**, *6*, 257–272. [CrossRef]
28. Zheng, H.W.; Shen, G.Q.; Wang, H. A review of recent studies on sustainable urban renewal. *Habitat Int.* **2014**, *41*, 272–279. [CrossRef]
29. Zheng, H.W.; Shen, G.Q.; Song, Y.; Sun, B.; Hong, J. Neighborhood sustainability in urban renewal: An assessment framework. *Environ. Plan. B Urban Anal. City Sci.* **2017**, *44*, 903–924. [CrossRef]
30. Huang, L.; Zheng, W.; Hong, J.; Liu, Y.; Liu, G. Paths and strategies for sustainable urban renewal at the neighbourhood level: A framework for decision-making. *Sustain. Cities Soc.* **2020**, *55*, 102074. [CrossRef]
31. Ho, D.C.W.; Yau, Y.; Poon, S.W.; Liusman, E. Achieving sustainable urban renewal in Hong Kong: Strategy for dilapidation assessment of high rises. *J. Urban Plan. Dev.* **2012**, *138*, 153–165. [CrossRef]
32. Tarani, P. Emergent creative ecosystems: Key elements for urban renewal strategies. In Proceedings of the 4th Knowledge Cities World Summit, Bento Goncalves, Brazil, 26 October 2011.
33. Van Melik, R.; Lawton, P. The role of public space in urban renewal strategies in Rotterdam and Dublin. *Plan. Pract. Res.* **2011**, *26*, 513–530. [CrossRef]
34. Miltenburg, E.M.; van de Werfhorst, H.G.; Musterd, S.; Tieskens, K. Consequences of forced residential relocation: Early impacts of urban renewal strategies on forced relocatees' housing opportunities and socioeconomic outcomes. *Hous. Policy Debate* **2018**, *28*, 609–634. [CrossRef]
35. Tsubari, A.; Alterman, R. *Agreements Between Planning Authorities and Developers: Tel Aviv's Innovative Policies for Exacting Public Amenities*; Center for Urban and Regional Studies books: Haifa, Israel, 2008.
36. *Public Infrastructure, Private Finance: Developer Obligations and Responsibilities*; Gielen, D.M.; van der Krabben, E. (Eds.) Routledge: Abingdon-on-Thames, UK, 2019.
37. Hendricks, A.; Auziņš, A.; Burinskienė, M.; Jürgenson, E. Vergleich der Abschöpfung entwicklungsbedingter Wertsteigerungen in Deutschland und dem Baltikum. *ZfV-Z. Geodäsie Geoinf. Landmanagement* **2018**, *143*, 242–249.
38. Botticini, F.; Auzins, A.; Lacoere, P.; Lewis, O.; Tiboni, M. Land Take and Value Capture: Towards More Efficient Land Use. *Sustainability* **2022**, *14*, 778. [CrossRef]
39. Roderick, Y.; McEwan, D.; Wheatley, C.; Alonso, C. Comparison of energy performance assessment between LEED, BREEAM and Green Star. In Proceedings of the Eleventh International IBPSA Conference, Glasgow, Scotland, 27–30 July 2009.
40. Lee, W.L.; Burnett, J. Benchmarking energy use assessment of HK-BEAM, BREEAM and LEED. *Build. Environ.* **2008**, *43*, 1882–1891. [CrossRef]
41. Sharifi, A.; Murayama, A. A critical review of seven selected neighborhood sustainability assessment tools. *Environ. Impact Assess. Rev.* **2013**, *38*, 73–87. [CrossRef]
42. Ameen, R.F.M.; Mourshed, M.; Li, H.A. critical review of environmental assessment tools for sustainable urban design. *Environ. Impact Assess. Rev.* **2015**, *55*, 110–125. [CrossRef]

43. U.S. Green Building Council. Leadership in Energy and Environmental Design (LEED). 2001. Available online: https://www.usgbc.org/leed (accessed on 29 March 2020).
44. Szibbo, N.A. Livability and LEED-ND: The Challenges and Successes of Sustainable Neighborhood Rating Systems. Ph.D. Thesis, UC Berkeley, Berkeley, CA, USA, 2015.
45. Wheeler, S.M. *Green Architecture and Building, Planning for Sustainability: Creating Livable, Equitable and Ecological Communities*, 2nd ed.; Routledge: Abingdon, UK, 2013; pp. 194–196.
46. Neighborhood 360. The Israeli Green Building Council (ILGBC). 2019. Available online: http://www.nd360.org/ (accessed on 5 December 2019).
47. Shach-Pinsly, D.; Capeluto, I.G. From Form-Based to Performance-Based Codes. *Sustainability* **2020**, *12*, 5657. [CrossRef]
48. Shach-Pinsly, D. Visual exposure and visual openness analysis model used as evaluation tool during the urban design development process. *J. Urban.* **2010**, *3*, 161–184. [CrossRef]
49. Frank, L.D.; Sallis, J.F.; Saelens, B.E.; Leary, L.; Cain, K.; Conway, T.L.; Hess, P.M. The development of a walkability index: Application to the neighbor- hood quality of life study. *Br. J. Sports Med.* **2010**, *44*, 924–933. [CrossRef]

Article

Mapping Resilience in the Town Camps of Mparntwe

Chris Tucker [1,*] , Michael Klerck [2] and Anna Flouris [2]

1   School of Architecture and Built Environment, University of Newcastle, Callaghan 2308, Australia
2   Tangentyere Council, Alice Springs 0870, Australia; michael.klerck@tangentyere.org.au (M.K.);
     anna.flouris@tangentyere.org.au (A.F.)
*   Correspondence: chris.tucker@newcastle.edu.au

Abstract: From the perspective of urban planning, the history of the Town Camps of Mparntwe (Alice Springs) has made them a unique form of urban development within Australia; they embody at once a First Nation form of urbanism and Country, colonial policies of inequity and dispossession, and a disparate public and community infrastructure that reflects the inadequate and ever-changing funding landscape it has been open to. While these issues continue, this paper discusses the resilience of these communities through the Local Decision Making agreement, signed in 2019 between the Northern Territory Government and Tangentyere Council. One thing that has been critical to translating and communicating local decisions for government funding has been the establishment of an inclusive and robust process of participatory mapping—*Mapping Local Decisions*—where both the deficiencies and potential of community infrastructure within each Town Camp is being identified. As local community knowledge is embedded within these practices, so too are issues of health, accessibility, safety and a changing climate similarly embedded within the architectural and infrastructure projects developed for government funding. Being conceived and supported by local communities, projects are finding better ways to secure this funding, building on a resilience these communities have for the places they live.

Keywords: Town Camps; First Nation communities; topological mapping; community infrastructure; PPGIS; minimalism; architectural design

Citation: Tucker, C.; Klerck, M.; Flouris, A. Mapping Resilience in the Town Camps of Mparntwe. *Architecture* **2022**, *2*, 446–456. https://doi.org/10.3390/architecture2030025

Academic Editor: Avi Friedman

Received: 30 May 2022
Accepted: 15 June 2022
Published: 22 June 2022

1. Introduction

We locally identify the process used to discuss, map and design community infrastructure within the Town Camp as *Mapping Local Decisions*. This method builds on previous qualitative research using public participation geographic information systems (PPGIS) that aims to utilise the rich text and synergies of dialogue [1] within communities. These place-based methodologies are uniquely positioned for this research [2]. English is often not a first language for the First Nation people who live in the Town Camps, with twelve local Aboriginal languages spoken in them. Meanwhile, drawing the landscape and telling its story is also an important cultural way to tell others about Country. Methods that graphically record the usual ways that individuals talk to each other in Town Camp communities maintain a vital feedback loop, ensuring that proposals are visually located in the places that they will directly affect and use symbols that can be clearly understood no matter an individual's literacy background.

The need for local decisions concerning community infrastructure to be robustly recorded also relates to the historically complex tenure of land within the 16 Town Camps of Alice Springs, which has allowed its community infrastructure to develop at a lower standard to that provided in the rest of Mparntwe. Subsequently, many residents live in unacceptable and unsafe conditions, with restricted access to adequate housing, health, education and employment opportunities [3] (pp. 116–119). The responsibility for funding this infrastructure and its maintenance reinforces an inequality that continues a history of exclusion that residents have fought against since the 1880s, when the First Nation people of Central Australia began to be dispossessed of their traditional lands. Since that time, the residents of Town Camps have resisted a colonialism that sought their removal and assimilation [4] (p. 19). In the 1970's, Town Camps began to actively assert their rights, forming a council in 1977—Tangentyere Council Aboriginal Corporation—where a movement for independence, control and self-determination began and continues to this day [5] (p. xii). Most of the work presently undertaken by Tangentyere Council is aligned with action on the social, environmental and behavioural determinants of health and wellbeing, delivering programs throughout Central Australia. Through this, *Mapping Local Decisions* has been made possible through the strong relationship Tangentyere Council has with the communities of the Town Camps, and their cultural awareness in being able to engage with the communities in constructive discussion.

2. Local Decision Making

The Local Decision Making (LDM) agreement between Tangentyere Council and the NTG, initiated in 2019 and signed in 2020 [6], prioritized self-determination and community control within Town Camps [7]. The agreement includes objectives to respect *"the long established and strong systems of Town Camp governance and leadership in the Alice Springs Town Camps...to document the commitment by the NT Government and TCAC to work together to implement LDM in the Alice Springs Town Camps...and to identify the services and priorities over which Town Campers wish to have control and for which they wish to have responsibility"* [6] (p. 2). This agreement was a significant shift in the processes used for how decisions about the provision of community infrastructure were to be made. It followed on from the Federal government's Northern Territory National Emergency Response, a 2009 joint funding program between the Australian Federal Government and the Northern Territory Government (NTG) called the Strategic Indigenous Housing and Infrastructure Program (SIHIP) [8]. SIHIP was a government-led and -controlled initiative established to design and construct community infrastructure in a range of indigenous communities in the Northern Territory. The Town Camps were included in this funding; however, most expenditure related to the pressing need for housing (133 of 199 Town Camp houses were upgraded) [9], with very little funding for community infrastructure [10] (p. 3). Some improvements to the road and community infrastructure of the larger Town Camps— Yarrenyty Arltere, Ewyenper Atwatye and Nyewente—were made; however, none of the work completed met the standards outlined in the Alice Springs Town Council Subdivision Guidelines [11]. As the Tangentyere Council response to the 2016 Inquiry into Housing Repairs and Maintenance on Town Camps noted, this failure *"...is supported by the fact that the Alice Springs Town Council is unprepared to deliver Municipal and Essential Services on any Town Camp."* [10] (p. 3).

Complicating the responsibility for community infrastructure is that the NTG identifies the land occupied by the Town Camps as 'Community Living' intended for *" ... temporary and permanent accommodation, and non-residential facilities for the social, cultural and recreational needs of residents"* [12] (p. 27). While this zoning appears to cover most activities within Town Camps, the current leasing agreements do not allow for them to be maintained by local government in similar ways to other residential areas of Mparntwe. The 'Community Living' zoning also prevents economic development activities, which limits the potential and self-determinism of Town Camps. The recent *Town Camps Reform Framework* recognises this limitation and appears to have the aspiration to reform this, allowing land owners *" ... to use and develop the land in line with community and resident aspirations"* [13] (p. 10). Public space in the Town Camp is highly valued, and for many residents the space outside of the house is treated as a *living room*; this recognises a relationship to Country, an externally-oriented lifestyle, and a requirement to accommodate long- and short-stay visitors [14]. The *wellness* of residents has also become an important consideration in Town Camps [15], and while issues of housing have a significant impact on wellness [16], a study to investigate the environmental determinants to health and wellbeing in remote communities by Chakraborty et al. [17] identified that aboriginal people living in remote areas of the Northern Territory are disproportionately disadvantaged, with inequalities not only limited to poverty, but also influenced by the broader social determinants of health, education, employment, skills development, technological innovation, transport and social support: *" … it is imperative to reduce structural inequities in society through a more equitable distribution of community infrastructure resources, income, goods, and services for the holistic health and wellbeing of its people"*. Residents, being the primary users of the community infrastructure, also identified the role that the extreme heat of Mparntwe plays in its utility. The community space of a Town Camp is necessarily outside and heavily reliant on structures that provide shade, trees and water. As the climate continues to get hotter and private power usage increases and becomes more costly, community centres are increasingly offering a place of last resort for the many Town Camp residents who suffer energy insecurity [18]. Facing some of the highest temperatures nationally, Town Camp communities are vulnerable to the effects of a warming climate. As Longden et al. point out, exposure to extreme temperatures is associated with a range of adverse health outcomes including death [18]. To provide some context, the 2004 report into Climate Change in the Northern Territory [19] noted that Alice Springs averaged 90 days over 35 °C and 17 days over 40 °C (in 2004). This report predicted that by 2030 these figures would increase to between 96–125 days over 35 °C and to between 21–43 days over 40 °C. The figures for 2018/19 have already surpassed these predictions [18].

3. Mapping Local Decision Making and Resilience

Participatory mapping has emerged as an important method to identify the values of a place [20]. Brown et al. found that the mapping of 'place values' includes land-use preferences and is generally stable over time, important characteristics of a robust *Making Local Decisions* methodology. Powell et al. similarly describe participatory mapping as being able to highlight and display the *"involuted relationships between self and place and the ways in which self and place are mutually constitutive and relational"* [21]. While mapping the phenomena shaping a person's individual experience of a place is a challenging concept [22], the LDM process is undertaken in discussion with the community as a group, with the resulting maps reflecting a *"collective rather than an individual outcome"* [23]. The *Making Local Decisions* methodology is outlined as follows:

- A high-resolution aerial photograph of the Town Camp is made from digital Nearmap images and printed out in large format, 1200 mm square. A small group of researchers, together with staff from Tangentyere Council, visit the Town Camp to initiate LDM discussions with the community. Meeting times often align with morning tea or lunch, with numbers and the make-up of the community group varying over the next hour or two;
- Depending on the availability of a community space, the aerial photograph is laid out on a large table in an inside or outside community space, allowing people to stand or sit around its edges (see Figure 1). The high-resolution of the image captures the smallest of details within the landscape, while also showing the broader organisation of the Town Camp and the roads and landscape that provide access to it. This is the only document brought in for the LDM discussion. None of the visiting group have clipboards, notebooks or any other equipment that differentiates them from the community;

Figure 1. A Town Camp community gathering for a LDM Mapping process.

- The aerial photograph of the Town Camp is the centre-piece of the discussion, and people begin to engage with it immediately; its large size is novel but its content is relatable and easily understood. Fingers begin to run over pathways, and in a local language, residents point, discuss, laugh and gesture about what it shows. People find their own houses and they begin to discuss and tell stories about how the Town Camp works;

- The LDM process is introduced and then residents lead the conversation in a local language and sometimes in English. Issues with community infrastructure and housing weave in and out of conversations in different ways as resident groupings change over time. The accessibility of the aerial image invites contributions from all members of the community, no matter their age, literacy or language group. Children in particular appear drawn to the image; playful and enthused, they want to know what everything is while pointing out as much as they know;

- The community are encouraged to mark the image with felt-tip pens, locating: the routes of informal roads, occasional camping areas, broken street lights, breaks in fences, places that flood when it rains, bike tracks, routes people take when walking to town, or at night, the lack of playground fencing near fast moving cars, places for speed humps and pedestrian crossings, the lack of a road kerb and concrete pathways, and the lack of asphalt on roads (see Figure 2). These many discussions are about issues that other Australians and other residents of Alice Springs never have to live with. Discussions also reveal access to public transport, how people wait without shelter in extreme summer temperatures, and how energy insecurity poses significant health risks;

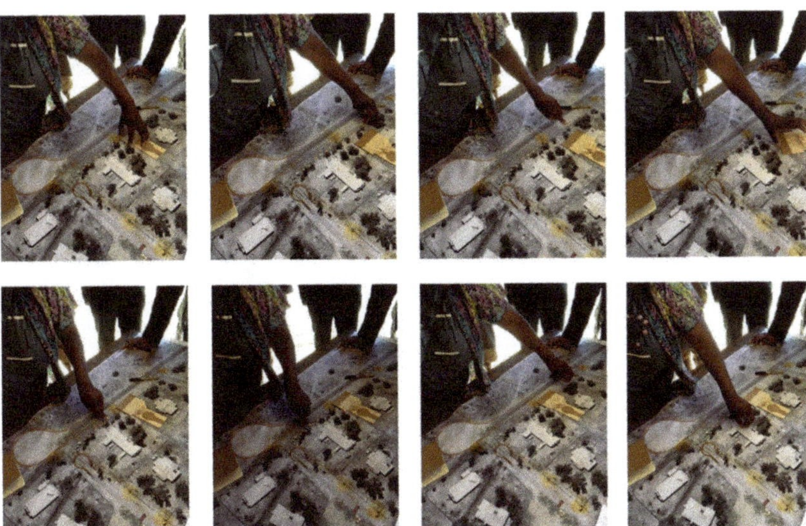

Figure 2. Examples of the LDM Mapping process from a Town Camp meeting.

- The aerial image becomes heavily marked in places with lines and symbols, disturbing its previously seamless qualities. We mobilise found objects from around the room and repurpose cardboard and timber blocks to represent buildings and structures. As the drawn marks and models grow in number, the aerial photograph gives way to the complexity of a topological map [24]. As the discussion highlights the past and future events of occupying the real space of the Town Camp, the visual qualities of the aerial photograph are transformed and differentiated to depict past events and issues and how the new projects will attempt to solve them. Each of these discrete projects is a topology, critically related and connected to others nearby. Now appearing on the map for all to see, apparent solutions to issues continue to be negotiated, edited and ultimately networked to each other as the mix of residents changes over time. *Mapping in this way is both a process and a tool for recording the conversations of Local Decision Making;*
- Following the meeting, the mark-ups left on the map are re-drawn digitally over the aerial image, with similar symbols being used as a record of the discussion. A few days later, the map with updated symbols is again printed out and a similar meeting is again made with the community to confirm, edit and add to what has been recorded. The map is again updated and refined to graphically depict Local Decisions as symbols. Models and more refined design drawings are also used where solutions will become community buildings or alterations to them (see Figures 3 and 4. A legend is now provided at the bottom of the map to confirm the meaning of the symbols to the community and for a broader audience that will follow (see Figure 5).
- When the Local Decisions within the map have been confirmed by the community, identified projects are tabulated in a schedule that both prioritises and itemises each element for costing. This schedule, together with the map, will form the basis of funding applications to territory and federal governments. As topologies, projects are also collated into the unpublished Guide to Infrastructure and Housing Standards for Town Camps (see Figure 6) so that similar issues and solutions between Town Camps can be identified and related to local, territory and national planning and regulatory requirements.

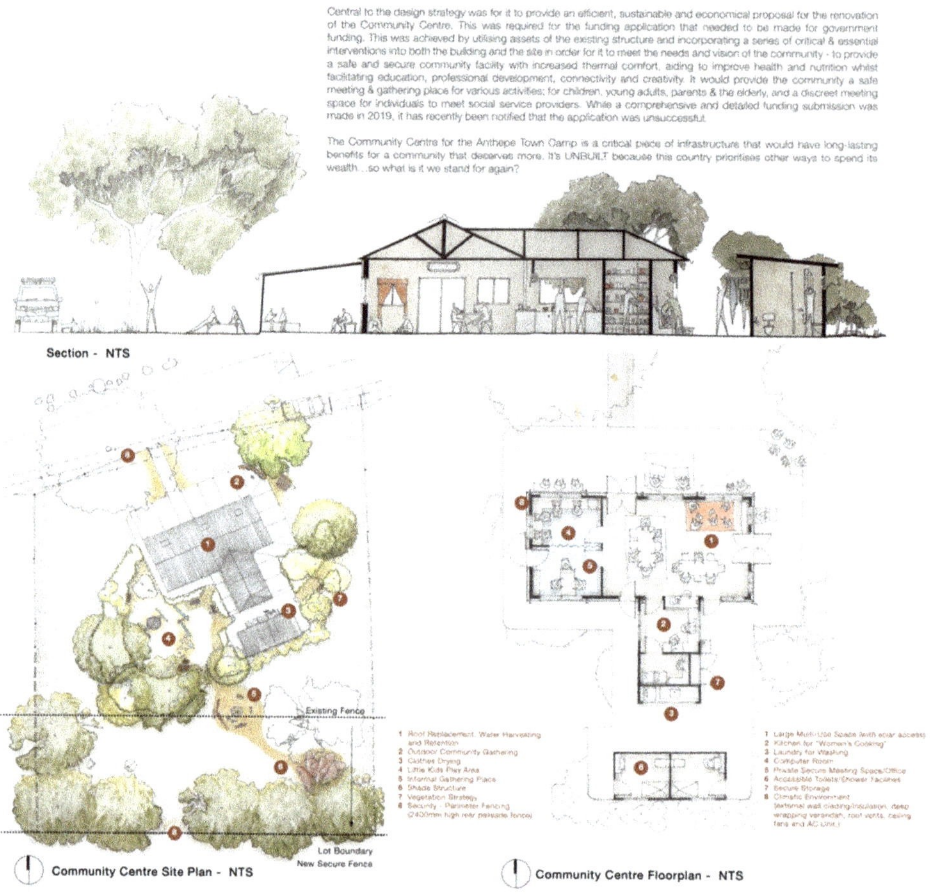
Central to the design strategy was for it to provide an efficient, sustainable and economical proposal for the renovation of the Community Centre. This was required for the funding application that needed to be made for government funding. This was achieved by utilising assets of the existing structure and incorporating a series of critical & essential interventions into both the building and the site in order for it to meet the needs and vision of the community - to provide a safe and secure community facility with increased thermal comfort, aiding to improve health and nutrition whilst facilitating education, professional development, connectivity and creativity. It would provide the community a safe meeting & gathering place for various activities; for children, young adults, parents & the elderly, and a discreet meeting space for individuals to meet social service providers. While a comprehensive and detailed funding submission was made in 2019, it has recently been notified that the application was unsuccessful.
The Community Centre for the Anthepe Town Camp is a critical piece of infrastructure that would have long-lasting benefits for a community that deserves more. It's UNBUILT because this country prioritises other ways to spend its wealth…so what is it we stand for again?
Section - NTS
Existing Fence
Lot Boundary
New Secure Fence
Community Centre Site Plan - NTS
Community Centre Floorplan - NTS

Figure 3. Proposed 2019 updates to the Anthepe Community Centre.

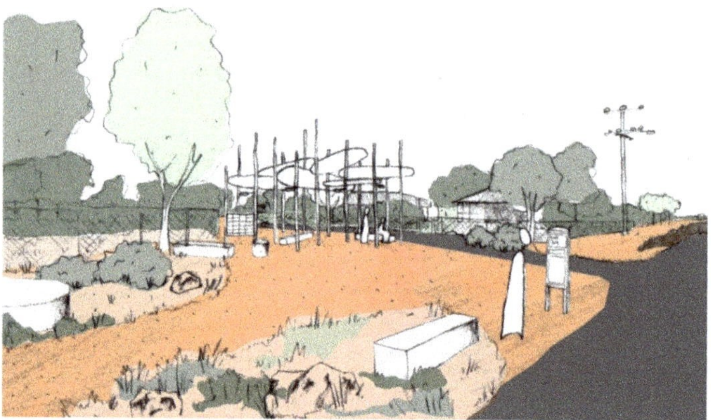

Figure 4. Sketch of shelter projects for Karnte.

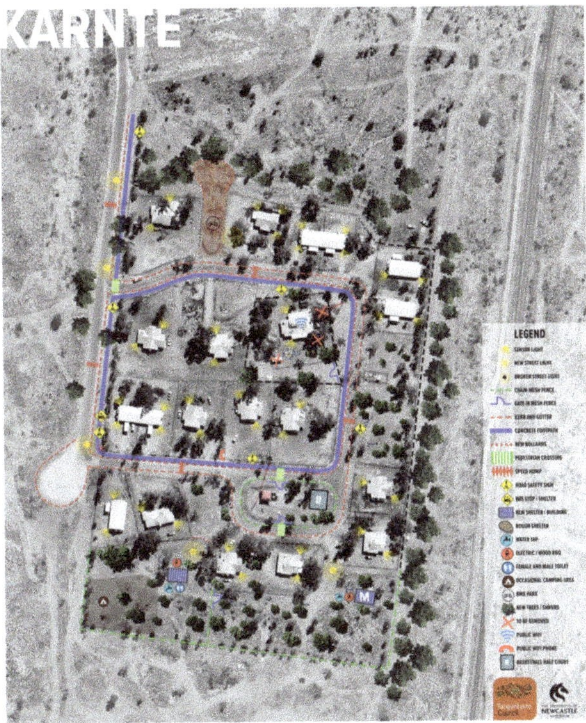

Figure 5. The LDM Map for Karnte.

Figure 6. An extract from the Guide to Infrastructure and Housing Standards for Town Camps showing the 'Roads' and 'Verge Planting' topology. The situation found in the Town Camp on the left is contrasted with the same situation typically found outside Town Camps in Alice Springs.

4. Discussion

In the last few years of utilising *Mapping Local Decisions*, it has shown itself to be a highly engaging way to stimulate discussion about the future of the Town Camp. As Brown et al. reflect, *"In the transference of human knowledge and experience to a map through PPGIS, humans are reminded of their identity and dependence on place … "* [23]. It invites a sharing of opinions that might otherwise not have been openly discussed and finds effective solutions to ongoing issues that may have never been found had a government-led *top-down* approach continued. The *Mapping Local Decisions* methodology, however, constantly evolves, with each Town Camp and each meeting providing greater insights into its process. The more time that is spent with the communities and engaged in *active listening* [25], the better we understand our role in translating local issues into architectural and community infrastructure proposals for government funding. The ongoing challenge of this research is to consciously work between two cultures [26], allowing conversations in the Town Camps to take their own shape, relying on the process of mapping to portray those discussions in accurate ways, and using more traditional architectural processes to communicate outcomes as funding proposals. The immediacy of working with the Town Camp community is vitally important to the research [27]; the non-verbal language [28], animation and position of those in the discussion is as important as anything that might actually be said. It allows relationships to be formed, sharpening an ability to apply architectural knowledge and develop skills that can quickly become useful.

In most towns and cities in Australia, communities don't need to engage with Local Decision Making and mapping processes to fund their community infrastructure. At some stage it was provided to them and remains essentially safe and well maintained; sealed roads have speed signs, stormwater drainage, kerbs and gutters, and there are footpaths, street lighting, pedestrian crossings and working community buildings. *The needs are so basic in Town Camps that it should embarrass a wealthy nation into immediate action; after so many years, why are we still talking about this?* The disparate state of community infrastructure within Town Camps reflects the colonial policies of inequity and dispossession, and the ever-changing funding landscape it has been subjected to [29]. As Senator Pat Dodson makes clear, the underlying issues of health, housing, education and employment need to be addressed as matters of urgency [30].

In terms of community infrastructure, the transition from suburban Mparntwe to the Town Camps is often stark. The usual asphalt roads become thinner, now with frayed edges as the kerb and gutter system disappears. With roads less defined, cars often leave the road entirely, creating a wider sandy zone that gets hollowed out and filled with water when it rains. This dusty edge optimises conditions for those that *drift* and travel at high speeds, making the side of the road a dangerous place for pedestrians. The ease with which cars can leave designated roads creates an informal road network that allows outsiders to enter the Town Camp in uncontrolled and dangerous ways. Those attempting to out-run the police also seek out the Town Camps, looking for these roads, knowing that their chase will be called off. At other times, police cars routinely prowl these informal roads for stolen cars and goods, keeping those who live nearby in a state of unease. Apart from the obvious solution of providing roads that meet minimum standards [31], with the required kerb and gutter [32], bollards can also be used to keep fast moving cars on the road and away from pedestrians. While road authorities and governments debate responsibility for these roads, none have the usual speed limit signs associated with suburban Mparntwe. More regular speed humps, designed to regulated standards, can also slow cars down. Stormwater needs to be collected at the side of the road as in nearby suburbs or placed in absorption trenches where it might help an already dry vegetated nature strip develop beside the road. Concrete pedestrian paths are almost universally non-existent in Town Camps and street lighting is a real problem, often well outside current regulation [33], located in ad-hoc and ineffective ways, and with bulbs not working. Where provided, community infrastructure is often so squeezed by the available funds that it is deployed in defensive, obligatory and loveless ways, detaching residents from its ownership and the full potential of community space. As Crabtree et al. point out, the Town Camps of Mparntwe make the case that *"Aboriginal communities remain caught up in an ongoing melee of political opportunism, ideological posturing, dubious contractual dealings, and policy disjuncture, very little of which reflects or respects community experience, knowledge, or aspirations"* [34]. This, in many respects, is the importance of *Mapping Local Decisions*; it translates Local Decisions into community projects for government funding, making it clear within those documents how the existing infrastructure fails to meet the regulations and standards government has set for itself.

5. Conclusions

The PPGIS tools and methods utilised within *Mapping Local Decisions* has supported the effective inclusion of the community in LDM, with government funding beginning to flow into identified projects and some already constructed. These include new bollards in Ewyenper-Atwatyeare, constructed works at the A2E (Access to Education) Brown Street Youth Centre and Inarlenge Community Centre, upgrades to community and public infrastructure at Karnte, Anthepe, Anthelk-Ewlpaye, Lhenpe Artnwe and Ilyperenye, and accessibility upgrades to houses. Many other projects are documented and many of course are awaiting funding. While local government in Australia usually accounts for the inception and design of these types of community facilities, in the Town Camps it is organisations such as Tangentyere Council and the University of Newcastle that have come together to produce this work. It is in terms of this obligation that we have perhaps found better ways to engage across this cultural divide and fulfil a future that these resilient communities hold and have held for many decades.

Since 2019, when LDM was initiated, all 16 Town Camps in Mparntwe have undertaken some form of *Mapping Local Decisions*, with some continuing to refine and update proposals as circumstances change; residents see more potential in the process and appreciate the significant role it plays. The importance of housing and community infrastructure on public and environmental health outcomes cannot be overlooked. At present, the emphasis in the Northern Territory through the National Partnership Agreement is on new houses and new bedrooms. From the perspective of Tangentyere Council, the built environment and community infrastructure of the Town Camps and Remote Communities is a priority. These places need community infrastructure that meets Australian Standards and local government development guidelines, and recognizes the resilience and authority that local decisions made in Town Camps have in shaping the future.

Author Contributions: Conceptualization, C.T.; methodology, C.T., M.K. and A.F. All authors have read and agreed to the published version of the manuscript.

Funding: This research received no external funding.

Institutional Review Board Statement: Not applicable.

Informed Consent Statement: Informed consent was obtained from all subjects involved in the study.

Data Availability Statement: Not applicable.

Conflicts of Interest: The authors declare no conflict of interest.

References

1. Lowery, D.R.; Morse, W.C. A Qualitative Method for Collecting Spatial Data on Important Places for Recreation, Livelihoods, and Ecological Meanings: Integrating Focus Groups with Public Participation Geographic Information Systems. *Soc. Nat. Resour.* **2013**, *26*, 1422–1437. [CrossRef]
2. Bąkowska-Waldmann, E.; Kaczmarek, T. The Use of PPGIS: Towards Reaching a Meaningful Public Participation in Spatial Planning. *Int. J. Geo-Inf.* **2021**, *9*, 581. [CrossRef]

3.   Centre For Appropriate Technology. *Housing Experience: Post Occupancy Evaluation of Alice Springs Town Camp Housing 2008–2011*; Centre for Appropriate Technology: Alice Springs, Australia, 2013; Available online: https://www.academia.edu/511173 0/Centre_for_Appropriate_Technology_2013_Housing_Experience_Post_Occupancy_Evaluation_of_Alice_Springs_Town_ Camp_Housing_2008_2011_Centre_for_Appropriate_Technology_Alice_Springs_NT (accessed on 24 May 2022).
4.   Wigley, J.; Wigley, B. Remote Conundrums: The Changing Role of Housing in Aboriginal Communities. In *Take 2: Housing Design in Indigenous Australia*; Memmott, P., Ed.; Royal Australian Institute of Architects: Red Hill, Australia, 2013; pp. 18–25.
5.   Coughlan, F. Aboriginal Town Camps and Tangentyere Council: The Battle for self-determination in Alice Springs. 1991. Available online: https://web.archive.org/web/20200416165800/https://www.tangentyere.org.au/publications/research_ reports/coughlan-aboriginal-town-camps-and-tangentyere-council.pdf (accessed on 14 June 2022).
6.   Northern Territory Government. *Local Decision Making Heads of Agreement Tangentyere Council Aboriginal Corporation and the Northern Territory Government*; Northern Territory Government: Palmerston, Australia, 2020.
7.   Northern Territory Government. Local Decision Making. Available online: https://ldm.nt.gov.au/ (accessed on 24 September 2019).
8.   Northern Territory Government; Australian Government. *Strategic Indigenous Housing and Infrastructure Program—Review of Program Performance*; Northern Territory Government and Australian Government: Darwin, Australia, 2009. Available online: https://www.dss.gov.au/sites/default/files/documents/05_2012/sihip_review.pdf (accessed on 14 May 2022).
9.   Northern Territory Government. *Inquiry into Housing Repairs and Maintenance on Town Camps*; Northern Territory Government: Palmerston, Australia, 2016. Available online: https://parliament.nt.gov.au/__data/assets/pdf_file/0008/366119/Terms_of_ Reference.pdf (accessed on 14 May 2022).
10. Tangentyere Council. *Submission No 11 into the Northern Territory Government Inquiry into Housing Repairs and Maintenance on Town CampsTangentyere Council Aboriginal Corporation*; Tangentyere Council: Alice Springs, Australia, 2016. Available online: https://parliament.nt.gov.au/__data/assets/pdf_file/0007/362491/Submission_No_11_Tangentyere_Council_Aboriginal_ Corporation.PDF (accessed on 15 May 2022).
11. Alice Springs Town Council. *Alice Springs Town Council Subdivision Development Requirements August 2018*; Alice Springs Town Council: Alice Springs, Australia, 2018; Available online: https://assets-astc.s3-ap-southeast-2.amazonaws.com/files/ files/documents/ASTC%20Subdivision%20Development%20Requirements%20August%202018%20-%20endorsed%20by%20 Council.pdf (accessed on 15 May 2022).
12. Northern Territory Government. *Northern Territory Planning Scheme—Part Four: Zones and Assessment Tables*; Northern Territory Government: Palmerston, Australia, 2020. Available online: https://nt.gov.au/__data/assets/pdf_file/0010/914869/nt-planning-scheme-part-four-zones-and-assessment-tables.pdf (accessed on 15 May 2022).
13. Northern Territory Government. *Town Camps Reform Framework 2019–2024*; Northern Territory Government: Palmerston, Australia, 2019. Available online: https://tfhc.nt.gov.au/__data/assets/pdf_file/0008/716660/town-camps-framework.pdf (accessed on 15 May 2022).
14. Memmott, P.; Remote Protoype. Architecture AU. 2001. Available online: https://architectureau.com/articles/remote-protoype/ (accessed on 17 May 2022).
15. Northern Territory Government; Tangentyere Council. *Town Camper Wellness Framework*; Northern Territory Government: Palmerston, Australia, 2019; Available online: https://www.tangentyere.org.au/town-camper-wellness-framework (accessed on 14 May 2022).
16. Pholeros, P.; Lea, T.; Rainow, S.; Sowerbutts, T.; Torzillo, P.J. Improving the state of health hardware in Australian Indigenous housing: Building more houses is not the only answer. *Int. J. Circumpol. Health* **2013**, *72*, 21181. [CrossRef] [PubMed]
17. Chakraborty, A.; Daniel, M.; Howard, N.J.; Chong, A.; Slavin, N.; Brown, A.; Cargo, M. Identifying Environmental Determinants Relevant to Health and Wellbeing in Remote Australian Indigenous Communities: A Scoping Review of Grey Literature. *Int. J. Environ. Res. Public Health* **2021**, *18*, 4167. [CrossRef] [PubMed]
18. Longden, T.; Quilty, S.; Riley, B.; White, L.V.; Klerck, M.; Davis, V.N.; Frank Jupurrurla, N. Energy insecurity during temperature extremes in remote Australia. *Nat. Energy* **2022**, *7*, 43–54. [CrossRef]
19. CSIRO. *Climate Change in the Northern Territory: State of the Science and Climate Change Impacts*; CSIRO: Perth, Australia, 2020.
20. Brown, G.; Reed, P.; Raymond, C.M. Mapping place values: 10 lessons from two decades of public participation GIS empirical research. *Appl. Geogr.* **2020**, *116*, 102156. [CrossRef]
21. Powell, K. Making Sense of Place: Mapping as a Multisensory Research Method. *Qual. Inq.* **2010**, *16*, 539–555. [CrossRef]
22. Bidwell, N.J.; Browning, D. Pursuing genius loci: Interaction design and natural places. *Pers. Ubiquitous Comput.* **2010**, *14*, 15–30. [CrossRef]
23. Brown, G.; Kyttä, M. Key issues and research priorities for public participation GIS (PPGIS): A synthesis based on empirical research. *Appl. Geogr.* **2014**, *46*, 122–136. [CrossRef]
24. Manning, E. Relationscapes: How Contemporary Aboriginal Art Moves Beyond the Map. *Cult. Stud. Rev.* **2007**, *13*, 134–155. [CrossRef]
25. Broffman, A. An architecture of listening. *Archit. Ausralia* **2008**, *97*, 90–96.
26. Gajendran, T.; Tucker, C.; Ware, S.; Tose, H. Integrating Indigenous, Western and inclusive pedagogies for work-integrated learning partnerships in architecture and design disciplines. *Int. J. Work. Integr. Learn.* **2022**, *23*, 259–277.

27. Peluso, N.L. Whose Woods are these? Counter-Mapping Forest Territories in Kalimantan, Indonesia. *Antipode* **1995**, *4*, 383–406. [CrossRef]
28. Yunkaporta, T. Aboriginal Pedagogies at the Cultural Interface. Ph.D. Thesis, James Cook University, Townsville, Australia, 2009.
29. Tangentyere Council Aboriginal Corporation. *Submission: To the Public Accounts Committee Inquiry into Local Decision-Making (LDM)*; Tangentyere Council Aboriginal Corporation: Alice Springs, Australia, 2021. Available online: https://dipl.nt.gov.au/__data/assets/pdf_file/0014/240404/Performance-%20and-Design-Standards-for-Northern-Territory-Government-Roads.pdf (accessed on 21 March 2021).
30. The Guardian. *Time to Stop the Rot: Pat Dodson Gives Fiery Speech about Indigenous Deaths in Custody*; The Guardian: London, UK, 2020.
31. Northern Territory Government. *Performance and Design Standards for Northern Territory Government Roads*; Northern Territory Government: Palmerston, Australia, 2017. Available online: https://dipl.nt.gov.au/__data/assets/pdf_file/0014/240404/Performance-and-Design-Standards-for-Northern-Territory-Government-Roads.pdf (accessed on 21 March 2021).
32. Alice Springs Town Council. *Developers Guide NOV 2012 Standard Drawings (ALL)*; Alice Springs Town Council: Alice Springs, Australia, 2012; Available online: https://assets-astc.s3-ap-southeast-2.amazonaws.com/files/files/documents/Developers%20Guide%20NOV%202012%20Standard%20Drawings%20(ALL).pdf (accessed on 21 March 2021).
33. Northern Territory Government. *Street Lighting Design Guidelines*; Northern Territory Government: Palmerston, Australia, 2020. Available online: https://dipl.nt.gov.au/__data/assets/pdf_file/0005/802760/street-lighting-design-guidelines.pdf (accessed on 21 March 2021).
34. Crabtree, L.; Davis, V.; Foster, D.; Klerck, M. A tale of (at least) three reports: Agency and discourse in the Alice Springs town camps. *Glob. Media J.* **2018**, *12*, 1.

MDPI

St. Alban-Anlage 66

4052 Basel

Switzerland

Tel. +41 61 683 77 34

Fax +41 61 302 89 18

www.mdpi.com

Architecture Editorial Office

E-mail: architecture@mdpi.com

www.mdpi.com/journal/architecture